THERMODYNAMIC VALUES
AT LOW TEMPERATURE
FOR NATURAL INORGANIC MATERIALS

THERMODYNAMIC VALUES AT LOW TEMPERATURE FOR NATURAL INORGANIC MATERIALS:

An Uncritical Summary

TERRI L. WOODS and ROBERT M. GARRELS

University of South Florida at St. Petersburg
Department of Marine Science

New York   Oxford

OXFORD UNIVERSITY PRESS

1987

Oxford University Press

Oxford   New York   Toronto
Delhi   Bombay   Calcutta   Madras   Karachi
Petaling Jaya   Singapore   Hong Kong   Tokyo
Nairobi   Dar es Salaam   Cape Town
Melbourne   Auckland

and associated companies in
Beirut   Berlin   Ibadan   Nicosia

Published by Oxford University Press, Inc.,
200 Madison Avenue, New York, New York 10016

Library of Congress Cataloging-in-Publication Data

Woods, Terri L.
Thermodynamic values at low temperature for
natural inorganic materials.

Includes bibliographical references.
1. Thermochemistry—Tables.   2. Inorganic compounds—
Thermal properties—Tables.   I. Garrels, Robert Minard,
1916–   II. Title.
QD511.8.W66   1987   541.3′6′0212   86-16442
ISBN 0-19-504888-1 (alk. paper)

1 3 5 7 9 10 8 6 4 2

Printed in the United States of America
on acid-free paper

Acknowledgment

Gratitude for valuable assistance in the preparation of the table is due to Joan Hutton, Judy Robertson, and Valerie Kimball.

CONTENTS

INTRODUCTION

This uncritical table of thermodynamic data gives scientists and engineers a comprehensive list of values from the major sources currently available, applicable to naturally occurring common elements and inorganic compounds. The table provides not only a quick reference list of values and sources of values but also gives an idea of the range of values determined for various compounds (i.e., of the degree of controversy over the best value for a given compound). An example of another uncritical summary of thermodynamic data, which has been widely used by scientists working in many fields, is *Stability Constants* by Sillen and Martel (1964).

The table is particularly useful for comparing the entropies or Gibbs energies of reactions, using data from different sources. Even though the Gibbs energies of formation of individual species may differ markedly from author to author, the Gibbs energy of the reaction may be the same if all data are from a single source.

Moreover, the perils of indiscriminate selection of values from various sources can be tested by calculating the Gibbs energies of a given reaction, using data chosen at random from the table.

The table contains values, where known, of the enthalpy of formation, Gibbs energy of formation, and entropy at 298.15°K and 1 bar pressure. Included are values for common elements, aqueous species, gases, liquids, and minerals and related compounds, as well as many minerals that are not found in critical tables. All of the energy values are expressed in terms of joules or kilojoules. Values reported in the literature in calories were converted to joules during compilation (*Conversion factor*: 1 calorie = 4.184 joules). If less than or equal to one decimal place was listed in the source, the original number of significant figures was retained. When more than one decimal place was listed in the original source, the values were rounded to one decimal place.

SOURCES OF DATA

Many critical evaluations of thermodynamic data have been used extensively to compile this work. The most comprehensive are the tables for organic and inorganic substances of the NBS (National Bureau of Standards, Wagman et al., 1982). These, however, do not include many of the common minerals of interest to earth scientists and engineers. The compilation by Robie et al. (1978) includes many common minerals but only a few aqueous species, concentrating on values for minerals at elevated temperatures. The compilations of Helgeson et al. (1978) and Naumov et al. (1971) appear to be the most internally consistent. Helgeson et al. (1978), however, include only a limited number of rock-forming minerals and no aqueous species. These values also frequently differ significantly from the average of values from other sources and often do not reflect observed mineral stabilities at low temperatures. The most recent Russian compilation draws on the earlier work by Naumov et al. (1971), indicating that the Russian consensus on most values has changed little in recent years.

STANDARD STATES

All values of thermodynamic properties apply at 298.15°K and 1 bar pressure. The standard states used are defined below:

1. *Pure solid*: Pure substance in the solid phase.
2. *Pure liquid*: Pure substance in the liquid phase.
3. *Gas*: Pure substance at a pressure of 1 bar in a "hypothetical" state in which it exhibits ideal gas behavior.
4. *Aqueous species*: Solute dissolved in hypothetical ideal aqueous solution with unit molality.

Some of the older values were determined for 1 atmosphere (1 atmosphere = 0.986923 bar). There are no cases in this list for which the pressure coefficients for condensed phases are large enough to affect the tabulated values. The pressure coefficients for gases are large enough to affect high-accuracy data. In practice the only properties tabulated here that are affected are $S°$ (for gases) and $\Delta G_f°$ for some substances (i.e., gases or substances for which one of the reference elements is a gas).

ARRANGEMENT OF SUBSTANCES IN THE TABLE

The data are presented in seven columns. The first contains the formula; the second, a description, if necessary; and the third, the physical state of the substance. The next three columns contain the enthalpy of formation, the Gibbs free energy of formation, and the entropy, respectively, for one mole of the substance. The seventh column, headed "Source," identifies the reference from which the number was taken. These references are listed both alphabetically and numerically following the table. The substances are arranged in the table by the first element in the formula, except for hydrogen, and the elements are arranged alphabetically. Within the listing for each element the compounds are arranged in the following order:

Element in its standard state
Element in nonstandard states

Aqueous species of the element
  Element itself
  Oxides
  Hydroxides
  Halides
  Carbonates
  Sulfates
  Ammonia-containing species
  Cyanide-containing species
Oxides
Hydroxides
Halides (listed in the order; Cl, F, Br, I, mixed anion halides)
Sulfides
Selenides
Tellurides
Hydrides
Carbides
Carbonates
Sulfates
Selenates
Phosphates
Nitrates
Arsenates
Manganates
Molybdates
Ferrites
Tungstates
Borates
Aluminates

Chromates
Uranates
Silicates
Titanates
Titano-silicates
Silicate-carbonates

PHYSICAL STATE

The physical state of each substance is indicated in the column headed "State" as crystalline solid (c), liquid (l), glassy or amorphous (am), gaseous (g) or aqueous species (aq). Our "aq" designation indicates that the value given is for the aqueous species occurring, as written, in solution [i.e., completely associated such as the bicarbonate ion (HCO_3^-) and not completely dissociated into the ions of which it is assumed to be composed at infinite dilution (in this case H^+ and CO_3^{2-})]. Except for the NBS tables (Wagman et al., 1982) the sources used to compile this table listed only one designation for aqueous species and only one value for each species. We have therefore assumed that their values are for the associated species.

Wagman et al. (1982) differentiate aqueous species as to whether they are associated, and with respect to the amount of solvent in which the substances are dissolved, if the solution is other than infinitely dilute. We have incorporated all NBS designations under the single designation "aq." We have not included any NBS values for substances in solutions for which a specific solution composition is given; (i.e., "in 1000 H_2O"), but we have included species designated by NBS as "aq," indicating a solute dissolved in dilute aqueous solution of unspecified composition. We have identified these species by noting in the description column that they are for an unspecified aqueous solution. For some species, the NBS table lists both an associated and a dissociated value. Except for one species in our table ($LiNO_3$ (aq)) the NBS dissociated value is equal to the sum of the values for the component parts of the species. For example,

$$\Delta G_f^\circ \ Cd(OH)_2(aq) =$$
$$\Delta G_f^\circ \ Cd^{2+}(aq) + 2\Delta G_f^\circ \ OH^-(aq).$$

Therefore, any NBS value that we list is for the associated species as written.

ABBREVIATIONS

The following abbreviations are used in the table:

deg	degree
est	estimated by RMG
J	joule
kJ	kilojoule
ppt.	precipitate
pptd.	precipitated
temp.	temperature
unspc. aq. soln.	unspecified aqueous solution

ATOMIC WEIGHTS FOR 1983

Element	Symbol	Atomic Weight	Element	Symbol	Atomic Weight	Element	Symbol	Atomic Weight
Actinium	Ac	227.0280	Holmium	Ho	164.930	Radium	Ra	226.025
Aluminum	Al	26.9815	Hydrogen	H	1.0080	Radon	Rn	(222)
Americium	Am	(243)	Indium	In	114.82	Rhenium	Re	186.2
Antimony	Sb	121.75	Iodine	I	126.9044	Rhodium	Rh	102.905
Argon	Ar	39.948	Iridium	Ir	192.2	Rubidium	Rb	85.4678
Arsenic	As	74.9216	Iron	Fe	55.847	Ruthenium	Ru	101.07
Astatine	At	(210)	Krypton	Kr	83.80	Samarium	Sm	150.35
Barium	Ba	137.34	Lanthanum	La	138.91	Scandium	Sc	44.956
Beryllium	Be	9.0122	Lead	Pb	207.19	Selenium	Se	78.96
Bismuth	Bi	208.980	Lithium	Li	6.941	Silicon	Si	28.086
Boron	B	10.811	Lutetium	Lu	174.97	Silver	Ag	107.870
Bromine	Br	79.909	Magnesium	Mg	24.312	Sodium	Na	22.9898
Cadmium	Cd	112.40	Manganese	Mn	54.9380	Strontium	Sr	87.62
Calcium	Ca	40.08	Mercury	Hg	200.59	Sulfur	S	32.064
Carbon	C	12.0112	Molybdenum	Mo	95.94	Tantalum	Ta	180.948
Cerium	Ce	140.12	Neodymium	Nd	144.24	Technetium	Tc	98.906
Cesium	Cs	132.9054	Neon	Ne	20.183	Tellurium	Te	127.60
Chlorine	Cl	35.453	Neptunium	Np	237.0482	Terbium	Tb	158.924
Chromium	Cr	51.996	Nickel	Ni	58.71	Thallium	Tl	204.37
Cobalt	Co	58.9332	Niobium	Nb	92.906	Thorium	Th	232.0381
Copper	Cu	63.54	Nitrogen	N	14.0067	Thulium	Tm	168.934
Dysprosium	Dy	162.50	Osmium	Os	190.2	Tin	Sn	118.69
Erbium	Er	167.26	Oxygen	O	15.9994	Titanium	Ti	47.90
Europium	Eu	151.96	Palladium	Pd	106.4	Tungsten	W	183.85
Fluorine	F	18.9984	Phosphorus	P	30.9738	Uranium	U	238.0290
Francium	Fr	223.0000	Platinum	Pt	195.09	Vanadium	V	50.942
Gadolinium	Gd	157.25	Plutonium	Pu	239.05	Xenon	Xe	131.30
Gallium	Ga	69.72	Polonium	Po	(209)	Ytterbium	Yb	173.04
Germanium	Ge	72.59	Potassium	K	39.1020	Yttrium	Y	88.905
Gold	Au	196.967	Praseodymium	Pr	140.907	Zinc	Zn	65.37
Hafnium	Hf	178.49	Promethium	Pm	146.915	Zirconium	Zr	91.22
Helium	He	4.0026	Protactinium	Pa	231.0359			

TABLE OF THERMODYNAMIC DATA

Formula	Description	State	ΔH°_f kJ	ΔG°_f kJ	S° J/deg	Source

- ALUMINUM -

Formula	Description	State	ΔH°_f kJ	ΔG°_f kJ	S° J/deg	Source
Al	Metal	c	0	0	28.3	3
			0	0	28.3	6
			0	0	28.3	11
			0	0	28.3	21
			0	0	28.3	31
Al^{3+}		aq	−524.7	−481.2	−313.4	3
			−531.4	−492.0		6
				−500.0		8
			−531.0	−489.4	−308.0	21
			−531.	−485.	−321.7	31
			−531.4	−489.4	−308.2	52
			−525.1	−489.5		71a
			−529.7	−483.7		71b
			−531.4	−492.0	−299.6	119
$Al(OH)^{2+}$		aq	−767.0	−700.6	−156.9	6
				−698.3		23

Formula	Description	State	ΔH°_f kJ	ΔG°_f kJ	S° J/deg	Source
				−693.9		29
				−694.1		31
			−760.6	−698.3		71a
			−765.2	−692.4		71b
				−689.4		89
			−767.0	−700.6	−156.9	119
$Al(OH)_2^+$		aq		−907.4		6
				−905.8		23
				−914.2		28
				−910.0		29
				−900.0		89
				−907.4		119
$Al(OH)_4^-$		aq		−1297.9		5
			−1490.3	−1305.4	144.3	6
				−1311.7		23
				−1300.2		27
				−1305.		28
				−1301.2		29

Formula	Description	State	ΔH_f° kJ	ΔG_f° kJ	S° J/deg	Source
			−1502.5	−1305.3	102.9	31
				−1304.2		57
			−1490.8	−1311.3		71a
			−1495.4	−1305.4		71b
				−1313.4		89
				−1301.3		90
			−1490.3	−1305.4		119
Al_2O_3	Corundum	c	−1669.8	−1576.4	51.0	3
			−1675.6	−1582.4	50.9	6
				−1582.0		10
			−1669.8	−1576.4	51.0	11
			−1675.7	−1582.2	50.9	21
			−1661.6	−1568.3	51.0	22
				−1581.9		27
			−1675.7	−1582.3	50.9	31
			−1661.6	−1568.3		58
				−1579.1		63
				−1581.9		99

Formula	Description	State	ΔH°_f kJ	ΔG°_f kJ	S° J/deg	Source
			−1675.7	−1582.2	50.9	121
$Al(OH)_3$	Gibbsite	c	−1294.5	−1156.9	70.1	6
				−1159.0		8
			−1279.0	−1141.4	70.1	11
				−1160.2	70.1	12
			−1293.1	−1154.9	68.4	21
			−1293.1	−1155.5	70.1	22
				−1154.8		23
			−1295.2	−1150.4	70.1	24
				−1148.4		27
				−1151.9		29
			−1293.3	−1155.1	68.4	31
			−1293.1	−1154.9	68.4	47
			−1293.1	−1154.9		71
				−1151.2		78
				−1159.4		89
				−1156.6		99
			−1294.6	−1156.5	68.4	108

Formula	Description	State	ΔH_f° kJ	ΔG_f° kJ	S° J/deg	Source
			-1293.3	-1155.1	68.4	121
$Al(OH)_3$	Bayerite	c	-1289.5			6
				-1155.8		12
				-1147.1		27
				-1153.		28
				-1148.9		29
			-1288.2			31
				-1146.4		57
$Al(OH)_3$		am	-1275.7	-1137.6	71.	1
			-1276.1	-1142.2	82.8	6
			-1276.			31
$Al(OH)_3$	Microcrystalline	c		-1139.3		57
$AlO(OH)$	Boehmite	c	-985.	-910.0	48.4	3
			-987.8	-913.2	48.4	6
			-982.6	-908.0	48.4	11
			-993.0	-918.4	48.4	21
			-983.6	-909.	48.4	22
				-920.9		23

Formula	Description	State	ΔH°_f kJ	ΔG°_f kJ	S° J/deg	Source
				-910.7		27
			-993.1	-918.4	48.4	28
				-915.0		29
			-990.4	-915.8	48.4	31
				-916.0		63
				-910.4		78
			-990.7	-916.1	48.4	108
			-990.4	-915.8	48.4	121
AlO(OH)	Diaspore	c		-910.2		2
			-1000.0	-921.3	35.2	6
			-993.3	-914.8	35.3	11
			-1000.6	-922.0	35.3	21
			-992.3	-913.8	25.3	22
				-918.2		27
			-1000.6	-922.0	35.3	28
				-918.4		29
			-999.4	-920.9	35.3	31
				-920.2		63

Formula	Description	State	ΔH°_f kJ	ΔG°_f kJ	S° J/deg	Source
				-914.6		78
				-920.1		99
					35.3	105
			-999.9	-921.4	35.3	108
			-999.4	-920.9	35.3	121
AlF_3		c	-1510.4	-1431.1	66.5	6
			-1510.4	-1431.1	66.5	21
			-1504.1	-1425.0	66.4	31
Al_2S_3		c	-508.8	-492.4	96.	3
			-723.4			6
			-724.			31
$Al_2(SO_4)_3$		c	-3435.0	-3091.9	239.3	3
			-3440.8	-3100.1	239.3	6
			-3440.8	-3099.8	239.3	21
			-3440.8	-3099.9	239.3	31
$AlPO_4$	Berlinite	c	-1733.8	-1601.2	90.8	4
			-1733.8	-1618.0	90.8	6
			-1733.8	-1623.3	90.8	21

Formula	Description	State	ΔH°_f kJ	ΔG°_f kJ	S° J/deg	Source
			-1733.8	-1617.9	90.8	31
$AlPO_4 \cdot 2H_2O$	Variscite	c	-2353.3	-2111.4	134.5	6
				-2087.0		45
$Al_2PO_4(OH)_3$	Augelite	c		-2765.2		45
$Al_3(PO_4)_2(OH)_3(H_2O)_4 \cdot H_2O$	Wavellite	c		-5564.7		45
Al_2SiO_5	Kyanite	c	-2594.4	-2451.4	92.2	6
			-2589.0	-2438.7	83.8	11
			-2591.7	-2441.3	83.8	21
			-2581.1	-2430.7	83.7	22
			-2594.3	-2443.9	83.8	31
				-2442.6		63
			-2596.0	-2445.1	82.3	66
				-2443.4		99
			-2596.3	-2445.6	82.4	108
			-2594.3	-2444.0	84.5	121
Al_2SiO_5	Sillimanite	c	-2589.0	-2442.4	96.2	6
			-2584.1	-2437.1	96.1	11
			-2585.8	-2439.0	96.1	21

Formula	Description	State	ΔH°_f kJ	ΔG°_f kJ	S° J/deg	Source
			−2573.6	−2427.1	96.8	22
			−2587.8	−2441.0	96.1	31
				−2438.6		63
			−2587.8	−2440.9	95.8	66
				−2441.8		99
			−2588.9	−2442.1	95.6	108
			−2587.8	−2441.0	96.1	121
Al_2SiO_5	Andalusite	c	−2592.2	−2446.4	93.2	6
			−2586.8	−2439.3	93.2	11
			−2587.5	−2439.9	93.2	21
			−2576.8	−2429.2	92.9	22
			−2590.3	−2442.7	93.2	31
				−2441.0		63
			−2591.9	−2443.7	91.4	66
				−2444.0		99
			−2592.6	−2444.6	91.5	108
			−2590.3	−2442.8	93.8	121
$Al_6Si_2O_{13}$	Mullite	c	−6819.5	−6436.2	254.4	6

Formula	Description	State	ΔH_f° kJ	ΔG_f° kJ	S° J/deg	Source
			−6804.1	−6426.1	269.6	11
			−6810.4	−6431.3	269.6	21
			−6816.2	−6432.7	255.	31
			−6826.7	−6443.3		119
$Al_2Si_2O_7$	Metakaolinite	c	−3313.7	−3112.5		11
			−3378.4	−3173.4	124.2	119
$Al_2Si_2O_5(OH)_4$	Kaolinite	c		−3700.7		2
			−4098.6	−3778.2	202.9	6
				−3807.4		8
			−4118.8	−3797.8	202.9	11
			−4117.7	−3801.2		15a
			−4110.6	−3789.9		15b
				−3807.4		16
			−4120.1	−3799.4	203.0	21
			−4109.6	−3789.1	203.0	22
			−4100.6	−3780.0	203.0	24
			−4119.6	−3799.7	205.0	31
			−4120.1	−3799.4	203.0	47

Formula	Description	State	ΔH_f° kJ	ΔG_f° kJ	S° J/deg	Source
			−4109.6	−3789.1	203.0	58
				−3782.3		78
				−3783.2		90
				−3802.0		99
			−4133.6	−3802.5	205.1	108
			−4119.8	−3799.6	205.0	121
$Al_2Si_2O_5(OH)_4$	Dickite	c	−4097.4	−3775.0	197.1	6
			−4118.8	−3796.3	197.1	21
			−4118.3	−3795.9	197.1	31
			−4118.8	−3796.3	197.1	47
				−3784.0		78
			−4132.3	−3798.8	197.1	108
			−4118.5	−3795.9	197.1	121
$Al_2Si_2O_5(OH)_4$	Halloysite	c	−4080.0	−3759.5	203.3	6
			−4101.5	−3780.7	203.0	21
			−4101.2	−3780.5	203.3	31
			−4101.5	−3780.8	203.3	47
				−3761.		78

Formula	Description	State	ΔH_f° kJ	ΔG_f° kJ	S° J/deg	Source
			-4114.8	-3783.2	203.3	108
			-4101.0	-3780.4	203.3	121
$Al_2Si_4O_{10}(OH)_2$	Pyrophyllite	c		-5271.8		8
			-5633.0	-5262.3		11
				-5271.8		16a
				-5242.6		16b
			-5643.3	-5269.4	239.4	21
			-5628.8	-5255.1	239.3	22
			-5642.0	-5268.1	239.4	31
				-5266.5		63
			-5629.2	-5261.0	256.2	73
			-5640.0	-5266.0	236.8	74
				-5266.4		78
				-5268.1		99
			-5642.3	-5268.6	239.3	108
			-5642.0	-5268.1	239.4	121
$Al_2SiO_4(OH)_2$	Hydroxyl-topaz	c		-2693.2	112.0	92
$Al_2SiO_4F_2$	Fluor-topaz	c	-3084.4	-2910.7		91

Formula	Description	State	ΔH_f° kJ	ΔG_f° kJ	S° J/deg	Source
					105.4	92
$(Al_7Mg)(Si_{14}Al_2)O_{40}(OH)_8$	Illite	c			1104.2	21
$(Al_{2.58}Mg_{0.89}Fe_{0.67}^{3+})$ $(Si_{7.64}Al_{0.36})O_{20}(OH)_4$	Aberdeen montmorillonite (not electrically neutral)	c		-10273.8		39
$(Al_{3.03}Mg_{0.58}Fe_{0.45}^{3+})$ $(Si_{7.87}Al_{0.13})O_{20}(OH)_4$	Belle Fourche/Clay Spur montmorillonite (not electrically neutral)	c		-10352.4		38
Al_2TiO_5		c			109.6	6
					109.6	21
					109.6	119

– ANTIMONY –

Formula	Description	State	ΔH_f° kJ	ΔG_f° kJ	S° J/deg	Source
Sb	Metal	c	0	0	43.9	3
			0	0	45.7	6
			0	0	45.7	11
			0	0	45.5	21
			0	0	45.7	31
Sb		g	254.4	213.8	180.2	3

Formula	Description	State	ΔH°_f kJ	ΔG°_f kJ	S° J/deg	Source
			262.3	222.1	180.3	31
Sb_2		g	218.	167.	254.8	3
			235.6	187.	254.9	31
SbO^+		aq		-175.7		3
				-177.1		31
SbO_2^+		aq		-274.0		12
SbO_2^-		aq		-345.2		1
				-340.2		31
SbO_3^-		aq		-514.3		12
$HSbO_2$		aq		-407.9		1
			-487.9	-407.5	46.4	31
SbS_2^-		aq		-54.4		1
$Sb_2S_4^{2-}$		aq	-109.6	-49.8	-26.2	31
SbS_3^{2-}		aq		-133.9		1
Sb_2O_4		c	-808.8	-694.1		1
				-694.1		12
			-907.5	-795.7	127.2	31
Sb_4O_6	Senarmonite, cubic	c	-1409.2	-1246.8	246.0	3

Formula	Description	State	ΔH_f° kJ	ΔG_f° kJ	S° J/deg	Source
			-1440.8	-1282.0	264.8	6
			-1440.6	-1268.1	220.9	31
Sb_4O_6	Valentinite, orthorhombic	c	-1417.5	-1263.6	282.1	6
				-1230.1		12
			-1417.1	-1252.7	246.0	21
			-1417.1	-1253.0	246.0	31
Sb_2O_5		c	-980.7	-838.9	125.1	3
			-1007.5	-864.7	125.1	6
			-971.9	-829.2	125.1	31
$SbCl_3$		c	-382.2	-324.8	186.2	3
			-382.2	-323.7	184.1	6
				-324.8		12
			-382.2	-323.7	184.1	31
$SbCl_3$		g	-314.6	-302.5	338.1	3
			-313.8	-301.2	338.5	6
			-313.8	-301.2	337.8	31
SbF_3		c	-908.8	-836.0	105.4	1
			-915.5			31

Formula	Description	State	ΔH°_f kJ	ΔG°_f kJ	S° J/deg	Source
SbH_3		g	142.	147.7	222.	1
			145.1	147.8	232.8	31
Sb_2S_3	Stibnite	c	−157.7	−156.1	182.0	6
			−174.9	−173.5	182.0	21
			−174.9	−173.6	182.0	31
Sb_2S_3		am	−150.6	−133.9	126.8	1
			−126.4			6
			−147.3			31

— ARSENIC —

Formula	Description	State	ΔH°_f kJ	ΔG°_f kJ	S° J/deg	Source
As α	Gray metal	c	0	0	35.1	3
			0	0	35.6	6
			0	0	35.1	11
			0	0	35.7	21
			0	0	35.1	31
As		am	4.2			1
			13.6			6

Arsenic

Formula	Description	State	ΔH°_f kJ	ΔG°_f kJ	S° J/deg	Source
	β		4.2			31
As γ	Yellow	c	14.8			3
			14.6			31
As		g	253.7	212.3	174.1	3
			302.5	261.0	174.2	31
As_2		g	123.8	73.2	239.7	3
			193.7	142.7	242.2	6
			222.2	171.9	239.4	31
As_4		g	149.4	105.4	289.	3
			143.7	87.9	329.7	6
			143.9	92.4	314.	31
AsO^+		aq		−163.6		3
				−156.5		6
				−163.8		31
				−156.5		119
AsO_2^-		aq		−350.2		3
			−429.0	−350.0	40.6	31
AsO_4^{3-}		aq	−870.3	−636.	−144.8	3

Formula	Description	State	ΔH_f° kJ	ΔG_f° kJ	S° J/deg	Source
			−892.4	−652.0	−164.8	6
			−888.1	−648.4	−162.8	31
			−892.4	−652.0	−164.8	119
$HAsO_4^{2-}$		aq	−898.7	−707.	3.8	3
			−910.6	−717.7	−5.4	6
			−906.3	−714.6	−1.7	31
			−910.6	−717.7	−5.4	119
$H_2AsO_3^-$		aq	−712.5	−587.4		1
			−721.0	−593.3	110.9	6
			−714.8	−587.1	110.5	31
			−689.1	−524.2	−15.1	119
$H_2AsO_4^-$		aq	−904.6	−748.5	117.	3
			−913.8	−757.5	117.2	6
			−909.6	−753.2	117.	31
			−913.8	−757.5	117.2	119
$H_3AsO_3^\circ$		aq	−741.8	−639.7	196.6	3
			−748.5	−646.0	195.2	6
			−742.2	−639.8	195.0	31

Formula	Description	State	ΔH°_f kJ	ΔG°_f kJ	S° J/deg	Source
			−748.5	−646.0	195.2	119
$H_3AsO_4^\circ$		aq	−898.7	−769.0	206.3	3
			−906.7	−770.0	182.8	6
			−902.5	−766.0	184.	31
			−906.7	−770.0	182.8	119
As_2O_5		c	−914.6	−772.4	105.4	3
			−924.7	−782.0	105.4	6
			−924.9	−782.3	105.4	31
As_2O_3 α	Arsenolite, octahedral	c	−656.8	−576.0	107.1	3
			−666.1	−588.3	116.7	6
			−657.0	−576.0	107.4	21
			−657.0	−576.2	107.1	31
As_2O_3 β	Claudetite, monoclinic	c	−664.0	−589.1	126.8	6
			−654.8	−575.6	113.3	21
			−654.8	−577.0	117.	31
AsO		g	20.0			3
			70.0			31
$As_2O_5 \cdot 4H_2O$		c	−2093.2	−1720.0	261.9	1

Formula	Description	State	ΔH°_f kJ	ΔG°_f kJ	S° J/deg	Source
			-2104.6			31
$AsCl_3$		g	-299.2	-286.6	327.2	3
			-271.1	-258.1	326.2	6
			-261.5	-248.9	327.2	31
AsF_3		g	-913.4	-898.3	289.0	3
			-920.6	-905.7	289.0	6
			-785.8	-770.8	289.1	31
AsH_3		g	171.5	175.7	217.6	1
			66.4	68.9	222.8	31
As_2S_2	Realgar	c	-133.5	-134.5	137.6	1
			-145.6	-140.6	127.0	6
			-142.7	-140.64	127.0	21
			-142.7			31
As_2S_3	Orpiment	c	-146.4	-135.8	112.1	1
			-96.2	-95.4	163.6	6
			-169.0	-168.4	163.6	21
			-169.0	-168.6	163.6	31
As_2S_2		g	-17.6			3

Formula	Description	State	ΔH°_f kJ	ΔG°_f kJ	S° J/deg	Source

- BARIUM -

Formula	Description	State	ΔH°_f kJ	ΔG°_f kJ	S° J/deg	Source
Ba	Metal	c	0	0	67.	3
			0	0	60.7	6
			0	0	66.9	11
			0	0	62.4	21
			0	0	62.8	31
Ba^{2+}		aq	−538.4	−560.6	13.	3
			−524.0	−547.5	8.8	6
			−537.6	−560.7	9.6	21
			−537.6	−560.8	9.6	31
			−524.0	−547.5	8.8	119
BaO		c	−558.1	−528.4	70.3	3
			−538.1	−510.4	70.3	6
			−548.1	−520.4	72.1	21
			−553.5	−525.1	70.4	31
BaO_2		c	−629.7	−568.2	65.7	1
			−634.3			31

Formula	Description	State	ΔH°_f kJ	ΔG°_f kJ	S° J/deg	Source
$BaO_2 \cdot H_2O$		c	−935.1	−815.9	105.0	1
			−930.1			31
$Ba(OH)_2$		c	−946.4	−856.5	95.0	1
			−944.7			31
				−857.2		42
			−946.4	−856.5		119
$Ba(OH)_2 \cdot 8H_2O$		c	−3345.1	−2789.9		1
			−3342.2	−2792.8	427.	31
$BaCl_2$		c	−860.1	−810.8	126.	1
			−845.0	−797.3	123.7	6
			−858.6	−810.4	123.7	31
BaF_2		c	−1200.4	−1140.1	96.6	1
			−1191.6	−1141.8	96.4	6
			−1207.1	−1156.8	96.4	31
			−1196.6		96.2	119
BaS		c	−443.5	−437.2	78.2	1
					78.2	6
			−460.	−456.	78.2	31

Formula	Description	State	ΔH_f° kJ	ΔG_f° kJ	S° J/deg	Source
$BaCO_3$	Witherite	c	−1218.8	−1138.9	112.1	3
			−1201.0	−1123.0	112.1	6
			−1210.8	−1132.2	112.1	21
			−1244.7	−1164.8	112.1	22
			−1216.3	−1137.6	112.1	31
$BaSO_4$	Barite	c	−1465.2	−1353.1	132.2	3
			−1457.4	−1347.0	132.2	6
			−1473.2	−1362.2	132.2	21
			−1473.2	−1362.2	132.2	22
			−1473.2	−1362.2	132.2	31
				−1353.0		119
$Ba_3(PO_4)_2$		c	−4175.6	−3951.4	356.0	1
$Ba_3(PO_4)_2$	Colloidal	am	−4092.			31
$Ba(NO_3)_2$	Nitrobarite	c	−991.8	−795.0	213.8	1
			−978.5	−783.7	213.8	6
			−992.1	−796.6	213.8	21
			−992.1	−796.6	213.8	31
$BaMnO_4$		c	−1180.	−1075.	154.0	1

Formula	Description	State	ΔH_f° kJ	ΔG_f° kJ	S° J/deg	Source
				-1119.2		31
$BaMoO_4$		c	-1564.0	-1461.5	161.9	1
			-1537.0	-1432.6	149.0	6
			-1548.	-1439.6	138.	31
			-1541.8	-1435.1	148.1	119
$BaSeO_4$		c	-1171.5	-1061.9	151.0	1
			-1145.2	-1031.8	133.0	6
			-1146.4	-1044.7	176.	31
$BaWO_4$		c	-1705.8	-1563.1	172.	1
			-1633.8	-1528.8	151.9	6
			-1703.			31
$BaSiO_3$		c	-1504.1	-1417.1	101.2	1
			-1608.1	-1525.5	109.6	6
			-1623.6	-1540.2	109.6	31
			-1616.7			47
			-1628.3	-1544.2	112.1	119
Ba_2SiO_4		c	-2256.8	-2145.6	176.1	6
			-2287.8	-2174.8	176.1	31

Formula	Description	State	ΔH_f° kJ	ΔG_f° kJ	S° J/deg	Source
			-2275.4			47
			-2297.1	-2183.3	182.0	119
$Ba_2Si_3O_8$		c	-4153.8	-3933.2	258.2	6
			-4184.8	-3963.0	258.2	31
			-4169.3			47
			-4194.3	-3971.7	266.1	119
$BaSi_2O_5$		c	-2532.6	-2396.2	153.1	6
			-2548.1	-2410.7	153.1	31
			-2539.7			47
			-2553.0	-2414.5	154.0	119
Ba_3SiO_5		c	-3037.2	-2893.4	252.7	11
			-2966.0	-2822.4	252.7	119
Ba_2SiO_3		c	-2078.6	-1969.0	194.1	1

Formula	Description	State	ΔH_f° kJ	ΔG_f° kJ	S° J/deg	Source
		– BERYLLIUM –				
Be	Metal	c	0	0	9.5	3
			0	0	9.5	6
			0	0	9.5	11
			0	0	9.5	21
			0	0	9.5	31
Be^{2+}		aq	–389.1	–356.5	–230.	1
			–403.8	–381.2	–196.6	6
			–383.0	–379.7	–130.0	21
			–382.8	–379.7	–129.7	31
			–403.8	–381.2	–196.6	119
BeO_2^{2-}		aq	–785.8	–649.8	–113.	1
			–790.8	–640.1	–159.	31
Be_2O^{2+}		aq		–912.		1
$Be_2O_3^{2-}$		aq		–1246.8		1
$Be_3(OH)_3^{3+}$		aq		–1804.1		6
				–1801.6		31

Formula	Description	State	ΔH°_f kJ	ΔG°_f kJ	S° J/deg	Source
				-1804.1		119
BeO	Bromellite, hexagonal	c	-610.9	-581.6	14.1	3
			-607.3	-578.1	14.1	6
			-609.4	-580.1	13.8	21
			-609.6	-580.3		31
			-598.7	-569.6	14.2	119
BeO β		c	-601.8	-573.3	16.5	21
Be(OH)$_2$ α		c	-907.1	-820.9	55.6	1
			-902.6	-816.2	55.2	6
			-902.5	-815.0	51.9	31
Be(OH)$_2$ β		c	-904.2	-818.0	55.6	1
			-905.7	-818.5	52.7	6
			-905.8	-817.5	50.	31
BeO·Be(OH)$_2$	Pptd.	c	-1532.2	-1414.	69.9	1
BeCl$_2$ α		c	-511.7	-467.8	85.8	1
			-490.4	-445.7	82.7	6
			-490.4	-445.6	82.7	31
			-511.7	-497.0	86.0	119

Formula	Description	State	ΔH_f° kJ	ΔG_f° kJ	S° J/deg	Source
$BeCl_2$ β		c	−484.1		75.8	6
			−495.8	−448.9	75.8	31
BeS		c	−233.9	−233.9	38.9	1
			−234.3			31
BeH		g	326.8	298.3	170.9	3
			316.3	285.8	176.7	31
$BeSO_4$ α		c	−1196.6	−1088.7	90.0	1
			−1205.5	−1093.9	77.9	6
			−1198.3	−1088.7	90.0	119
$BeSO_4$	Tetragonal	c	−1205.2	−1093.8	77.9	31
$BeAl_2O_4$	Chrysoberyl	c	−2299.5	−2177.4	66.3	6
			−2300.8	−2178.5	66.3	21
			−2300.8	−2178.5	66.3	31
Be_2SiO_4	Phenacite	c	−2144.7	−2030.5	64.3	6
			−2158.7	−2044.3	64.3	11
					64.3	21
			−2149.3	−2032.5	64.3	31
			−2158.7	−2043.6	64.4	119

Formula	Description	State	ΔH°_f kJ	ΔG°_f kJ	S° J/deg	Source
$BeSiO_3$		c	−1539.6	−1455.6	54.0	11
			−1539.6	−1452.5	54.0	119

— BISMUTH —

Formula	Description	State	ΔH°_f kJ	ΔG°_f kJ	S° J/deg	Source
Bi	Metal	c	0	0	56.9	3
			0	0	56.9	6
			0	0	56.7	11
			0	0	56.7	21
			0	0	56.7	31
Bi		g	207.9	169.0	186.9	3
			207.1	168.2	187.0	31
Bi_2		g	248.5	200.8	273.6	3
			219.7			31
Bi^{3+}		aq	80.6	91.8	−176.6	6
				62.0		12
				82.8		21
				82.8		31

Formula	Description	State	ΔH°_f kJ	ΔG°_f kJ	S° J/deg	Source
			80.6	91.8	-176.6	119
BiO^+		aq		-144.5		3
				-146.0		6
				-146.4		31
$Bi(OH)^{2+}$		aq		-136.4		6
				-163.7		12
				-146.4		31
				-136.4		119
$BiCl_4^-$		aq		-477.8		1
				-479.5		6
				-481.5		31
				-479.5		119
BiO		c	-208.6	-182.0	71.	1
Bi_2O_3	Bismite	c	-577.0	-496.6	151.5	3
			-573.9	-493.7	151.5	6
			-573.9	-493.4	151.5	21
			-573.9	-493.7	151.5	31
Bi_2O_4		c		-456.0		1

Formula	Description	State	ΔH_f° kJ	ΔG_f° kJ	S° J/deg	Source
Bi_2O_5		c		-383.1		12
Bi_4O_7		c		-973.8		12
$Bi(OH)_3$		c	-714.6	-582.8	118.0	6
			-711.3			31
$Bi(OH)_3$		am	-709.6	-573.2	102.9	1
				-573.		12
$BiOOH$		c		-369.9		1
				-368.1		31
$BiCl$		g	44.8	21.8	246.4	3
$BiCl_3$		c	-379.1	-318.9	189.5	1
			-378.6	-313.0	172.	6
			-379.1	-315.0	177.0	31
$BiCl_3$		g	-270.7	-260.2	356.9	3
			-263.0	-252.5	356.5	6
			-265.7	-256.0	358.8	31
$BiOCl$	Bismoclite	c	-365.3	-322.2	86.2	3
			-369.4	-319.3	120.9	6
			-366.9	-322.1	120.5	31

Formula	Description	State	ΔH_f° kJ	ΔG_f° kJ	S° J/deg	Source
BiBr		g	53.1	15.9	257.7	3
BiI		g	67.	46.	265.3	3
Bi_2S_3	Bismuthinite	c	−183.2	−164.8	147.7	3
			−155.6	−153.1	200.4	6
			−143.1	−140.6	200.4	21
			−143.1	−140.6	200.4	31
BiH_3		g		231.5		12

– BORON –

Formula	Description	State	ΔH_f° kJ	ΔG_f° kJ	S° J/deg	Source
B		c	0	0	6.5	3
			0	0	5.9	6
			0	0	5.9	11
			0	0	5.9	21
			0	0	5.9	31
BO_2^-		aq	−767.8	−709.6	84.	1
			−772.4	−678.9	−37.2	31
BO_3^{3-}		aq		−759.3		12

Formula	Description	State	ΔH°_f kJ	ΔG°_f kJ	S° J/deg	Source
$B_4O_7^{2-}$		aq		−2577.3		1
				−2604.8		31
$B(OH)_4^-$		aq	−1343.1	−1153.1	105.0	6
			−1344.0	−1153.2	102.5	31
			−1343.1	−1154.7	105.0	119
$HB_4O_7^-$		aq		−2628.7		12
HBO_3^{2-}		aq		−838.0		12
$H_2BO_3^-$		aq	−1053.5	−910.6	30.5	3
$H_3BO_3^\circ$		aq	−1071.3	−968.6	164.8	6
				−963.0		12
			−1072.3	−968.8	162.3	31
			−1071.3	−968.6	164.8	119
BF_4^-		aq	−1527.	−1435.	167.	3
			−1573.7	−1491.4	200.4	6
			−1574.9	−1486.9	180.	31
			−1573.7	−1491.4	200.4	119
B_2O_3		c	−1263.6	−1184.1	54.0	3
			−1272.4	−1193.3	54.0	6

Formula	Description	State	ΔH_f° kJ	ΔG_f° kJ	S° J/deg	Source
			−1273.5	−1194.3	54.0	21
			−1272.8	−1193.6	54.0	31
B_2O_3		am	−1245.2	−1173.2	78.6	3
			−1254.2	−1182.0	77.8	6
			−1254.5	−1182.3	77.8	31
H_3BO_3		c	−1088.7	−963.2	89.6	3
			−1094.3	−968.9	88.8	31
BO		g		−81.7		12
			25.	−4.	203.5	31
BF_3		g	−1110.4	−1093.3	254.0	3
			−1135.3	−1118.6	254.0	6
			−1137.0	−1120.3	254.1	31
B_2S_3		c	−238.5	−223.0	57.3	1
			−240.6	−238.1	106.3	6
			−240.6			31
BH		g		471.1		12
			449.6	419.6	171.9	31
B_2H_6		g	31.4	82.8	232.9	3

Formula	Description	State	ΔH_f° kJ	ΔG_f° kJ	S° J/deg	Source
			35.6	86.7	232.1	31
B_5H_9		g	62.8	165.7	275.6	3
			73.2	175.0	275.9	31
$B_{10}H_{14}$		c		272.0		12
			−45.2	192.3	176.6	31
$B_{10}H_{14}$		g		297.		12
			31.6	216.3	353.2	31

– BROMINE –

Formula	Description	State	ΔH_f° kJ	ΔG_f° kJ	S° J/deg	Source
Br_2		l	0	0	152.3	3
			0	0	152.2	6
			0	0	152.2	11
			0	0	152.3	21
			0	0	152.2	31
Br_2	Ideal gas	g	30.7	3.1	245.3	3
			30.9	3.1	245.4	6
			30.9	3.1	245.5	21

Formula	Description	State	ΔH°_f kJ	ΔG°_f kJ	S° J/deg	Source
			30.9	3.1	245.5	31
Br^-		aq	-120.9	-102.8	80.7	3
			-121.5	-104.2	83.3	6
			-121.5	-104.0	82.8	21
			-121.6	-104.0	82.4	31
			-121.5	-104.2	83.3	119
Br_2°		aq	-4.6	4.1		1
			-0.8	4.0	136.0	6
			-2.6	3.9	130.5	31
			-0.8	4.0	136.0	119
Br_3^-		aq	-133.9	-105.7	167.4	1
			-132.2			6
			-130.4	-107.0	215.5	31
			-132.2			119
BrO^-		aq		-33.5		1
			-94.1	-33.4	42.	31
BrO_3^-		aq	-64.6	20.2	164.0	6
				9.6		12

Formula	Description	State	ΔH_f° kJ	ΔG_f° kJ	S° J/deg	Source
			−67.1	18.6	161.7	31
			−64.6	20.2	164.0	119
HBrO$^\circ$		aq		−83.3		1
			−113.0	−82.4	142.	31
HBr		g	−36.2	−53.2	198.5	3
			−36.4	−53.4	198.6	6
			−36.4	−53.4	198.7	31

- CADMIUM -

Formula	Description	State	ΔH_f° kJ	ΔG_f° kJ	S° J/deg	Source
Cd α	Metal	c	0	0	51.5	3
			0	0	51.8	6
			0	0	51.8	11
			0	0	51.8	21
			−0.6	−0.6	51.8	31
Cd γ	Metal	c		0.6		3
			0	0	51.8	31
Cd^{2+}		aq	−72.4	−77.7	−61.1	3

Formula	Description	State	ΔH_f° kJ	ΔG_f° kJ	S° J/deg	Source
			−75.6	−77.9	−71.1	6
			−75.9	−77.6	−73.2	21
			−75.9	−77.6	−73.2	31
			−75.9	−77.6		50
			−75.9	−77.7	−72.8	80
			−75.6	−77.9	−71.1	119
$HCdO_2^-$		aq		−361.9		12
				−363.5		31
$Cd(OH)^+$		aq		−271.5		6
				−261.1		31
				−271.5		119
$Cd(OH)_2$		aq		−446.4		6
				−442.6		31
				−446.4		119
$Cd(OH)_3^-$		aq		−605.0		6
				−600.7		31
				−605.0		119
$Cd(OH)_4^{2-}$		aq		−759.8		6

Formula	Description	State	ΔH°_f kJ	ΔG°_f kJ	S° J/deg	Source
				-758.4		31
				-759.8		119
$CdCl^+$		aq		-216.7	23.4	1
			-240.2	-220.5	31.8	6
			-240.6	-224.4	43.5	31
$CdCl^\circ_2$		aq		-352.7	71.	1
			-404.8	-355.6	109.6	6
			-405.0	-359.3	121.8	31
			-404.8	-355.6	109.6	119
$CdCl^-_3$		aq		-484.9	212.1	1
			-560.6	-483.7	193.3	6
			-561.1	-487.0	202.9	31
$Cd(NH_3)_4^{2+}$		aq		-224.8		1
			-450.2	-226.1	336.4	31
$Cd(CN)_4^{2-}$		aq		464.		1
			428.0	507.6	322.	31
CdO	Monteponite, cubic	c	-254.6	-225.0	54.8	3
			-258.0	-228.4	54.8	6

Formula	Description	State	ΔH_f° kJ	ΔG_f° kJ	S° J/deg	Source
			−258.2	−228.5	54.8	21
			−258.2	−228.4	54.8	31
$Cd(OH)_2$	"Active"	c	−557.6	−470.5	95.4	3
				−470.0		12
$Cd(OH)_2$	Pptd.	c	−560.7	−473.6	96.	31
$Cd(OH)_2$		c	−563.7	−474.5	87.9	6
$Cd(OH)_2$	"Inactive"	c		−473.3		12
			−560.6	−474.0		50
$CdCl_2$		c	−389.1	−342.6	118.4	3
			−391.7	−344.2	115.3	6
			−391.5	−343.9	115.3	31
CdF_2		c	−689.9	−647.7	113.	3
			−700.4	−651.1	89.1	6
			−700.4	−647.7	77.4	31
$CdBr_2$		c	−314.4	−293.5	133.5	3
			−316.3	−296.9	138.8	6
			−316.2	−296.3	137.2	31
CdS	Greenockite	c	−144.3	−140.6	71.	3

Formula	Description	State	ΔH°_f kJ	ΔG°_f kJ	S° J/deg	Source
			−151.9	−147.3	68.2	6
			−149.6	−145.6	70.3	21
			−161.9	−156.5	64.9	31
			−149.4	−145.2		50
CdTe		c	−101.7	−99.7	94.6	3
			−101.0	−98.2	92.0	6
			−92.5	−92.0	100.	31
CdH		g	261.7	233.2	212.4	3
$CdCO_3$	Otavite	c	−747.7	−670.3	105.4	3
			−754.0	−674.2	97.5	6
			−750.6	−669.4	92.5	21
			−750.6	−669.4	92.5	31
			−750.6	−669.4		50
$CdSO_4$		c	−926.2	−820.0	137.2	3
			−932.9	−822.4	123.0	6
			−933.3	−822.7	123.0	31
$CdSO_4 \cdot H_2O$		c	−1231.6	−1066.2	172.0	3
			−1239.2	−1068.5	154.0	6

Formula	Description	State	ΔH°_f kJ	ΔG°_f kJ	S° J/deg	Source
			−1239.6	−1068.7	154.0	31
$CdSO_4 \cdot 8/3H_2O$		c	−1723.0	−1462.8	242.2	3
			−1729.0	−1464.9	229.6	6
			−1729.4	−1465.1	229.6	31
$CdSiO_3$		c	−1189.0	−1105.3	97.5	6
			−1189.1	−1105.4	97.5	31
			−1185.5	−1101.8	97.5	47

- CALCIUM -

Formula	Description	State	ΔH°_f kJ	ΔG°_f kJ	S° J/deg	Source
Ca	Metal	c	0	0	41.7	6
			0	0	41.6	11
			0	0	41.6	21
			0	0	41.4	31
Ca^{2+}		aq	−543.0	−553.0	−55.2	3
			−542.7	−552.7	−55.2	6
				−553.1		8
			−542.8	−553.5	−53.1	21

Formula	Description	State	ΔH°_f kJ	ΔG°_f kJ	S° J/deg	Source
			−542.8	−553.6	−53.1	31
			−543.0	−552.8		50
			−542.7	−552.7	−55.2	119
$Ca(OH)^+$		aq		−717.8		2
			−764.3	−717.0	−14.6	6
				−718.4		31
			−764.3	−717.0	−14.6	119
$CaCO_3^o$		aq		−1099.4		2
				−1098.9		6
				−1098.9		119
$CaHCO_3^+$		aq		−1145.0		2
$CaSO_4^o$		aq		−1308.2		2
			−1448.5	−1310.4	20.5	6
			−1448.5	−1310.4	20.5	119
CaO	Lime	c	−635.5	−604.2	39.7	3
			−635.1	−603.9	39.7	6
			−635.7	−604.6	39.7	11
			−635.1	−603.5	38.2	21

Formula	Description	State	ΔH_f° kJ	ΔG_f° kJ	S° J/deg	Source
			−635.1	−604.0	39.7	22
			−635.1	−604.0	39.8	31
			−635.1	−603.5	38.1	121
CaO_2		c	−659.0	598.	43.1	1
			−652.7			31
$Ca(OH)_2$	Portlandite	c	−986.6	−896.3	76.1	3
			−984.6	−897.0	83.4	6
					83.4	11
			−986.1	−898.4	83.4	21
			−986.1	−898.5	83.4	31
				−897.6		72
$CaCl_2$	Hydrophilite	c	−795.0	−750.2	113.8	3
			−794.1	−749.1	113.8	6
			−795.8	−748.1	104.6	21
			−795.8	−748.1	104.6	31
CaF_2	Fluorite	c	−1214.6	−1161.9	68.9	3
			−1228.3	−1176.0	68.9	6
			−1229.3	−1176.9	68.9	21

Formula	Description	State	ΔH°_f kJ	ΔG°_f kJ	S° J/deg	Source
			-1225.9	-1173.6	68.6	22
			-1219.6	-1167.3	68.9	31
CaBr$_2$		c	-674.9	-656.0	130.	3
			-683.0			6
			-682.8	-663.6	130.	31
CaI$_2$		c	-534.7	-530.0	142.	3
			-537.5	-533.7	145.3	6
			-533.5	-528.9	142.	31
CaS	Oldhamite	c	-482.4	-477.4	56.5	3
					56.5	6
			-474.9	-470.0	56.6	21
			-482.4	-477.4	56.5	31
CaCO$_3$	Calcite	c	-1206.9	-1128.8	92.9	3
			-1206.8	-1128.3	91.7	6
			-1207.7		92.9	11
			-1207.4	-1128.8	91.7	21
			-1208.2	-1130.1	92.7	22
			-1206.9	-1128.8	92.9	31

Calcium

Formula	Description	State	ΔH°_f kJ	ΔG°_f kJ	S° J/deg	Source
			−1207.4	−1128.8		50
				−1129.3		72
				−1130.3		76
			−1209.0	−1130.6	91.8	108
$CaCO_3$	Aragonite	c	−1207.0	−1127.7	88.7	3
			−1207.0	−1127.4	88.0	6
				−1128.5	88.7	11
			−1207.4	−1127.8	88.0	21
			−1208.0	−1129.2	90.2	22
			−1207.1	−1127.8	88.7	31
				−1128.2		72
				−1129.4		76
			−1209.7	−1130.1	87.9	108
$CaCO_3$	Vaterite, metastable, hexagonal form	c		−1125.5		21
$CaCO_3 \cdot H_2O$	Monohydrocalcite	c	−1498.3	−1361.6		21
$CaMg(CO_3)_2$	Dolomite	c	−2331.7	−2169.3		3
			−2314.6	−2151.9	155.2	6

Formula	Description	State	ΔH°_f kJ	ΔG°_f kJ	S° J/deg	Source
			-2332.7	-2170.0	155.2	11
			-2324.5	-2161.7	155.2	21
			-2329.9	-2167.2	155.2	22
			-2326.3	-2163.4	155.2	31
				-2177.8		65
$CaMg(CO_3)_2$	Disordered Dolomite	c	-2317.6	-2158.4	166.7	22
$CaMg_3(CO_3)_4$	Huntite	c	-4529.6	-4203.4	299.5	21
				-4216.2		65
$CaMn(CO_3)_2$	Kutnahorite	c		-1950.6		65
$CaBa(CO_3)_2$	Barytocalcite	c		-2271.5		65
$CaBa(CO_3)_2$	Alstonite	c		-2271.9		65
$CaSO_4$	Anhydrite	c	-1432.7	-1320.3	106.7	3
			-1434.2	-1321.8	106.7	6
			-1436.4	-1324.1	106.7	11
			-1434.1	-1321.7	106.7	21
			-1434.1	-1321.8	106.7	22
			-1434.1	-1321.0	106.7	31
				-1323.0		72

Formula	Description	State	ΔH°_f kJ	ΔG°_f kJ	S° J/deg	Source
$CaSO_4$ α	Soluble	c	-1423.7	-1311.8	108.4	3
			-1425.2	-1313.4	108.4	31
$CaSO_4$ β	Soluble	c	-1419.2	-1307.3	108.4	3
			-1420.8	-1309.0	108.4	31
$CaSO_4 \cdot 0.5H_2O$ α	Bassanite	c	-1575.2	-1435.2	130.5	3
			-1576.6	-1461.3	130.5	6
	Macrocrystalline		-1576.7	-1436.7	130.5	31
$CaSO_4 \cdot 0.5H_2O$ β		c	-1573.0	-1434.2	134.3	3
			-1574.5	-1460.3	134.3	6
	Microcrystalline		-1574.6	-1435.8	134.3	31
$CaSO_4 \cdot 2H_2O$	Gypsum	c	-2021.1	-1795.7	194.0	3
			-2022.5	-1797.2	194.1	6
			-2022.6	-1797.2	194.1	21
			-2022.6	-1797.3	194.1	31
				-1798.4		72
$Ca_3(PO_4)_2$ α	High temp. form	c	-4126.3	-3889.9	241.0	3
			-4104.5	-3870.2	241.0	6
			-4109.9	-3875.5	240.9	31

Formula	Description	State	ΔH°_f kJ	ΔG°_f kJ	S° J/deg	Source
$Ca_3(PO_4)_2$ β	Low temp. form	c	−4137.6	−3899.5	236.0	3
			−4125.4	−3889.4	236.0	6
			−4120.8	−3895.6	236.0	21
			−4120.8	−3884.7	236.0	31
$CaHPO_4$	Monetite	c	−1820.9	−1679.9	88.	3
			−1813.2	−1680.0	111.4	6
			−1814.4	−1681.2	111.4	31
$Ca(H_2PO_4)_2$	Pptd.	c	−3114.6	−2812.	189.5	1
$Ca(H_2PO_4)_2$		c	−3104.1	−2824.8		4
			−3113.6			6
			−3104.7			31
$CaHPO_4 \cdot 2H_2O$	Brushite	c	−2410.0	−2153.1	167.4	3
			−2402.7	−2153.8	189.4	6
			−2403.6	−2154.6	189.4	31
$Ca_{10}(PO_4)_6(OH)_2$	Hydroxyapatite	c	−13482.9	−12684.2	780.7	6
			−13443.2	−12676.9	780.7	21
			−13477.	−12677.	780.7	31
$Ca_{10}(PO_4)_6(Cl_2)$	Chloroapatite, synthetic	c	−13272.0	−12514.0	795.8	4

Formula	Description	State	ΔH_f° kJ	ΔG_f° kJ	S° J/deg	Source
$Ca_{10}(PO_4)_6F_2$	Fluorapatite	c	−13775.4	−13014.8	775.7	6
			−13744.4	−13016.2	775.7	21
			−13744.	−12983.	775.7	31
$Ca_{9.54}Na_{0.33}Mg_{0.13}(PO_4)_{4.8}$ $(CO_3)_{1.2}F_{2.48}$	Carbonate-fluorapatite	c		−12294.5		4
$Ca_{10}(PO_4)_5CO_3F_2$	Carbonate-fluorapatite	c		−12482.1		45
$CaAl_3(PO_4)_2(OH)_5 \cdot H_2O$	Crandallite	c		−5592.8		45
$Ca_{1.5}Al_6(PO_4)_4(OH)_9 \cdot 3H_2O$	Ca-millisite	c		−10937.0		45
$Ca(NO_3)_2$		c	−938.0	−742.7	193.3	6
			−938.4	−743.0	193.3	21
			−938.4	−743.1	193.3	31
$CaMoO_4$	Powellite	c	−1546.0	−1439.3	122.6	6
			−1541.4	−1434.6	122.6	21
			−1541.4	−1434.6	122.6	31
			−1557.7	−1450.2	119.7	119
$CaFe_2O_4$		c	−1520.8	−1412.7	145.2	6
			−1520.3	−1412.7	145.4	21
			−1520.3	−1412.8	145.4	31

Formula	Description	State	ΔH°_f kJ	ΔG°_f kJ	S° J/deg	Source
$Ca_2Fe_2O_5$		c			188.7	6
			−2139.3	−2001.6	188.8	21
			−2139.3	−2001.7	188.8	31
			−2124.2	−1984.9	188.7	119
$CaWO_4$	Scheelite	c	−1642.2	−1542.6	151.0	1
			−1641.0	−1534.3	126.4	6
			−1645.2	−1538.4	126.4	21
			−1645.2	−1538.4	126.4	31
			−1596.2	−1493.7	140.2	119
$Ca_2B_6O_{11} \cdot 13H_2O$	Inyoite	c	−9293.			31
$Ca(UO_2)_2(PO_4)_2 \cdot 10H_2O$	Autunite	c		−7147.0		4
$CaSiO_3$	Wollastonite	c	−1584.1	−1498.7	82.0	3
			−1634.9	−1549.6	82.0	6
			−1631.1	−1545.6	82.0	11
			−1635.2	−1550.0	82.0	21
			−1631.0	−1545.8	82.0	22
			−1634.9	−1549.7	81.9	31
			−1635.2	−1549.9	82.0	47

Formula	Description	State	ΔH°_f kJ	ΔG°_f kJ	S° J/deg	Source
					86.1	53
			-1634.6	-1549.2	81.5	108
			-1634.8	-1549.2	81.0	121
$CaSiO_3$	β-Wollastonite	c	-1635.7	-1550.4	82.0	119
$CaSiO_3$	Pseudo-wollastonite	c	-1579.0	-1495.4	87.4	3
			-1631.6	-1547.9	87.4	6
			-1624.5	-1540.9	87.4	11
			-1628.6	-1545.0	87.4	21
			-1628.4	-1544.7	87.4	31
					86.4	53
			-1628.5	-1544.8	87.3	108
			-1627.6	-1543.9	87.2	121
$CaSiO_3$	α-Wollastonite	c	-1630.7	-1547.0	87.4	119
Ca_2SiO_4	Bredigite	c	-2310.2	-2192.2	116.6	108
Ca_2SiO_4	Ca-olivine	c	-2316.6	-2199.8	120.5	21
			-2316.6	-2199.8	120.5	47
					158.8	53
			-2317.5	-2200.8	120.5	108

Formula	Description	State	ΔH°_f kJ	ΔG°_f kJ	S° J/deg	Source
			−2316.5	−2199.7	120.5	121
Ca_2SiO_4 γ	Bredigite	c	−2317.9	−2201.2	120.5	6
Ca_2SiO_4 γ		c	−2255.2	−2149.3	157.3	1
			−2319.0	−2202.2	120.5	11
			−2317.9	−2201.1	120.8	31
			−2312.6	−2197.4	120.5	119
Ca_2SiO_4 β	Larnite	c	−2251.0	−2145.1	157.3	1
			−2307.4	−2192.8	127.6	6
			−2308.3	−2193.7	127.6	11
			−2306.0	−2191.3	127.6	21
			−2307.5	−2192.8	127.7	31
			−2306.0	−2191.3	127.6	47
					126.2	53
			−2307.2	−2192.4	127.1	108
			−2308.5	−2193.2	127.6	119
			−2306.7	−2191.7	126.7	121
$Ca_3Si_2O_7$	Rankinite	c	−3956.3	−3756.7	210.9	6
			−3957.2	−3757.6	210.9	11

Formula	Description	State	ΔH°_f kJ	ΔG°_f kJ	S° J/deg	Source
			-3961.0	-3761.3	210.8	31
			-3975.0	-3775.4	210.5	108
			-3927.6	-3727.1	210.9	119
			-3985.4	-3773.5	210.6	121
Ca_3SiO_5	Hatrurite	c	-2929.2	-2784.0	168.6	31
			-2933.7	-2788.3	168.6	108
			-2930.6	-2784.3	168.6	119
$CaAl_2Si_2O_8$	Anorthite	c	-4223.3	-3998.6	202.5	6
			-4217.9	-3993.0	202.5	11
			-4243.0	-4017.3	199.3	21
			-4216.5	-3992.8	205.4	22
			-4227.9	-4002.3	199.3	31
			-4243.0	-4017.3	199.3	47
				-4022.9		99
			-4230.7	-4005.0	199.3	108
			-4227.8	-4002.1	199.3	121
$CaAl_2Si_2O_8$	Anorthite, hexagonal	c	-4202.0	-3975.4	191.6	6
			-4197.4	-3968.9	191.8	11

Formula	Description	State	ΔH°_f kJ	ΔG°_f kJ	S° J/deg	Source
			−4222.6	−4001.4	214.8	21
			−4209.5	−3988.5	214.6	31
			−4222.6	−4001.4	214.8	47
$CaAl_2Si_2O_8$		am	−4151.4			6
			−4171.3	−3956.8	237.3	21
			−4155.1	−3940.8	237.2	31
			−4171.3	−3956.8	237.3	47
$CaAl_2SiO_6$	Ca−Al pyroxene	c	−3287.3	−3112.1		11
			−3280.2	−3108.3	156.0	21
			−3298.2	−3122.0	141.4	31
				−3142		99
			−3298.6	−3122.9	142.8	108
			−3299.0	−3122.8	141.6	121
$Ca_3Al_2Si_3O_{12}$	Grossular	c	−6577.2	−6235.4	322.0	11
			−6656.7	−6294.9	255.5	21
			−6624.9	−6263.3	254.7	22
					250.8	53
				−6314		99

Formula	Description	State	ΔH_f° kJ	ΔG_f° kJ	S° J/deg	Source
			−6638.9	−6277.6	256.3	108
			−6636.3	−6274.7	256.0	121
$Ca_2Al_2SiO_7$	Gehlenite	c	−3986.9	−3784.7	198.3	6
			−3981.9	−3779.6	198.3	11
			−4007.6	−3808.7	209.8	21
			−3981.8	−3780.6	201.2	22
			−3981.5	−3783.1	210.0	31
			−4007.6	−3808.7	209.8	47
					210.1	67
				−3808.7		99
			−3983.3	−3784.6	210.0	108
			−3981.7	−3782.9	209.9	121
$Ca_2Al_3Si_3O_{12}(OH)$	Clinozoisite	c	−6879.4	−6483.9	295.6	22
$Ca_2Al_3Si_3O_{12}(OH)$	Zoisite	c	−6852.9	−6464.0	318.6	11
			−6879.0	−6483.6	296.0	22
				−6535		99
			−6894.2	−6498.6	295.9	108
			−6891.1	−6495.3	295.9	121

Formula	Description	State	ΔH_f° kJ	ΔG_f° kJ	S° J/deg	Source
$Ca_2Al_2Si_3O_{10}(OH)_2$	Prehnite	c	-6181.9	-5803.8	289.1	11
			-6201.1	-5818.0	272.0	22
				-5837		99
			-6196.8	-5819.8	292.8	108
			-6193.6	-5816.4	292.7	121
$Ca_2Al_2SiO_6(OH)_2$	Bicchulite	c	-4337.2	-4071.2	217.3	108
$CaAl_4Si_2O_{10}(OH)_2$	Margarite	c	-6217.5	-5834.0	266.9	22
			-6242.9	-5858.4	263.7	108
			-6239.6	-5854.8	263.6	121
$CaAl_2Si_2O_7(OH)_2 \cdot H_2O$	Lawsonite	c	-4859.3	-4506.2	237.6	6
			-4853.5	-4500.3	237.6	11
			-4879.1	-4525.6	237.6	21
			-4846.4	-4492.0	233.5	22
			-4858.5	-4505.1	237.7	31
			-4879.1	-4525.6	237.6	47
				-4528.3		99
			-4875.2	-4522.1	175.2	119
$CaAl_2Si_4O_{12} \cdot 2H_2O$	Wairakite	c	-6637.4	-6198.3	397.0	11

Formula	Description	State	ΔH_f° kJ	ΔG_f° kJ	S° J/deg	Source
			−6608.8	−6182.5	439.7	22
				−6205		99
$CaAl_2Si_3O_{10} \cdot 3H_2O$	Scolecite	c	−6049.0	−5597.9	367.4	59
$CaAl_2Si_4O_{12} \cdot 4H_2O$	Laumontite	c	−7237.8	−6688.8	495.0	11
			−7233.6	−6682.0	485.8	22
			−7250.9	−6697.6	481.4	77
				−6707		99
$Ca_2Al_4Si_8O_{24} \cdot 7H_2O$	Leonhardite	c	−14217.6	−13169.1	922.2	6
			−14205.5	−13155.8	922.2	11
			−14246.5	−13197.1	922.2	21
			−14212.3	−13162.0	922.6	24
			−14249.0	−13199.8	922.9	31
			−14246.5	−13197.1	922.2	47
$CaAl_2Si_7O_{18} \cdot 6H_2O$	Heulandite	c	−10575.3	−9754.7	721.6	11
					763.2	22
$(Ca_{0.585}Ba_{0.065}Sr_{0.175}K_{0.132}$ $Na_{0.383})Al_{2.165}Si_{6.835}O_{18} \cdot 6H_2O$	Heulandite	c	−10491.0	−9675.6	767.2	118
$Ca_{0.167}Al_{2.33}Si_{3.67}O_{10}(OH)_2$	Ca-montmorillonite	c	−5723.6	−5352.3	256.1	11

Formula	Description	State	ΔH_f° kJ	ΔG_f° kJ	S° J/deg	Source
			-5726.4	-5354.9	256.2	24
			-5725.7	-5349.1	238.2	120
$Ca_3Fe_2Si_3O_{12}$	Andradite	c	-5742.8	-5403.2	327.0	11
			-5778.1	-5428.6	293.4	22
$CaFeSi_2O_6$	Hedenbergite	c	-2844.7	-2676.9	159.0	11
			-2838.8	-2674.5	170.3	22
			-2837.7	-2677.4		24
$Ca_2Fe_5Si_8O_{22}(OH)_2$	Actinolite	c	-10523.7	-9841.1	671.5	11
				-10985.3		24
$Ca_2Fe_3^{3+}Si_3O_{12}(OH)$	Epidote	c		-5205.7		24
$Ca_2Fe^{3+}Al_2Si_3O_{12}(OH)$	Epidote	c	-6461.9	-6072.4	315.0	22
$CaMgSi_2O_6$	Diopside	c	-3206.2	-3032.0	143.1	6
			-3204.2	-3029.8	143.1	11
			-3210.8	-3036.6	143.1	21
			-3203.3	-3029.2	143.1	22
			-3206.2	-3032.0	142.9	31
					140.8	53
$CaMgSiO_4$	Monticellite	c	-2263.2			6

Formula	Description	State	ΔH_f° kJ	ΔG_f° kJ	S° J/deg	Source
			-2257.8	-2139.9	108.4	11
			-2262.7	-2143.2	102.5	21
			-2262.7	-2145.7	110.4	22
			-2263.1			31
$Ca_2MgSi_2O_7$	Akermanite	c	-3877.3	-3679.8	209.2	6
			-3868.2	-3670.8	209.2	11
			-3876.5	-3679.1	209.3	21
			-3878.3	-3681.1	209.3	22
			-3877.2	-3679.8	209.2	31
$Ca_3Mg(SiO_4)_2$	Merwinite	c	-4567.5	-4340.2	253.1	6
			-4551.1	-4324.0	253.1	11
			-4566.8	-4339.4	253.1	21
			-4566.6	-4339.6	253.1	22
			-4567.7	-4340.3	253.1	31
$Ca_2Mg_5Si_8O_{22}(OH)_2$	Tremolite	c	-12358.3	-11612.3	548.9	6
			-12353.3	-11624.8	548.9	11
			-12355.1	-11627.9	548.9	21
			-12319.7	-11592.5	548.9	22

Formula	Description	State	ΔH_f° kJ	ΔG_f° kJ	S° J/deg	Source
			-12360.	-11631.	548.9	31
$Ca_4MgAl_5(H_2O)_2Si_6O_{23}(OH)_3$	Pumpellyite	c	-14239.8	-13363.6	757.6	11
$(Ca_{0.19}Na_{0.02}K_{0.02})(Al_{1.52}Fe_{0.14}^{3+}$	Cheto ca-montmorillonite	c		-5238.8		46
$Mg_{0.33})(Si_{3.93}Al_{0.07})O_{10}(OH)_2$				-5332.1		78
$CaTiO_3$	Perovskite	c	-1661.9	-1576.5	93.8	6
			-1660.6	-1575.2	93.6	21
			-1660.6	-1575.2	93.6	31
			-1660.6	-1575.3		119
$CaTiSiO_5$	Sphene	c	-2602.9	-2461.5	129.3	6
			-2604.3	-2460.8	129.3	11
			-2601.4	-2459.8	129.2	21
			-2603.3	-2461.8	129.2	31
			-2559.0	-2457.5	129.2	80
			-2568.8	-2427.3	129.3	119
$Ca_5(SiO_4)_2(CO_3)$	Spurrite	c	-5899.3	-5568.8	270.8	11
$Ca_5Si_2O_7(CO_3)_2$	Tilleyite	c	-6371.6	-6009.8	376.7	11
$Ca_4Al_6Si_6O_{24}(CO_3)$	Meionite	c	-13897.5	-13142.7	691.4	108

Formula	Description	State	ΔH_f° kJ	ΔG_f° kJ	S° J/deg	Source
		– CARBON –				
C	Graphite	c	0	0	5.7	3
			0	0	5.7	6
			0	0	5.4	7
			0	0	5.7	11
			0	0	5.7	21
			0	0	5.7	22
			0	0	5.7	31
C	Diamond	c	1.9	2.9	2.4	3
			1.9	2.9	2.4	6
			1.9	2.9	2.4	11
			1.9	2.9	2.4	21
			1.9	2.9	2.4	31
CO_2		aq	−412.9	−386.2	121.3	3
			−413.8	−386.0	120.9	6
			−413.8	−386.0	117.6	31
			−413.8	−386.0	120.9	119

Formula	Description	State	ΔH°_f kJ	ΔG°_f kJ	S° J/deg	Source
CO_3^{2-}		aq	−676.2	−528.1	−53.1	3
			−677.1	−527.9	−56.9	6
			−677.1	−527.9	−56.9	21
			−677.1	−527.8	−56.9	31
			−677.1	−527.9		50
			−677.1	−527.9	−56.9	119
$C_2O_4^{2-}$		aq	−818.8	−666.9	44.4	1
			−808.3	−667.6	79.9	6
			−808.3	−667.6	79.9	119
HCO_2^-		aq	−410.0	−334.7	91.6	3
			−425.6	−351.0	92.	31
$HC_2O_4^-$		aq	−818.8	−690.9		1
$H_2C_2O_4^\circ$		aq	−818.3	−697.9		1
HCO_3^-		aq	−691.1	−587.0	95.0	3
			−692.0	−586.8	91.2	6
			−692.0	−586.8	91.2	21
			−692.0	−586.8	91.2	31
			−692.0	−586.8		50

Formula	Description	State	ΔH°_f kJ	ΔG°_f kJ	S° J/deg	Source
			−692.0	−586.8	91.2	119
$H_2CO_3^\circ$		aq	−698.7	−623.4	191.2	3
			−699.6	−623.2	187.4	6
			−699.6	−623.2	187.0	21
			−699.6	−623.1	187.4	31
			−699.6	−623.2		50
			−699.6	−623.2	187.4	119
HCHO		aq		−129.7		1
			−141.8			31
HCO_2H		aq	−410.0	−356.0	163.6	3
			−425.4	−372.3	163.	31
CH_3OH		aq	−245.9	−175.2	132.2	3
			−245.9	−175.3	133.1	31
CH_4		aq	−89.0	−34.4	83.7	6
			−89.0	−34.3	83.7	31
			−89.0	−34.4	83.7	119
HCN		aq	107.1	119.7	124.7	31
CN^-		aq	150.6	172.4	94.1	31

Formula	Description	State	ΔH_f° kJ	ΔG_f° kJ	S° J/deg	Source
CH_3COO^-		aq	−486.0	−369.3	86.6	31
CO		g	−110.5	−137.3	197.9	3
			−110.5	−137.2	197.6	6
			−110.5	−137.2	197.7	21
			−110.5	−137.2	197.7	31
			−110.6	−137.2	−197.5	108
CO_2		g	−393.5	−394.4	213.6	3
			−393.5	−394.4	213.6	6
			−393.5	−394.4	213.7	11
			−393.5	−394.4	213.8	21
			−393.5	−394.4	213.7	22
			−393.5	−394.4	213.7	31
			−393.5	−394.4		50
CF_4		g	−679.9	−635.1	262.3	3
			−932.8	−888.2	261.5	6
			−925.	−879.	261.6	31
CS_2		g	115.3	65.1	237.8	3
			117.4	67.1	237.8	31

Formula	Description	State	ΔH°_f kJ	ΔG°_f kJ	S° J/deg	Source
C_2H_2		g	226.7	20.	200.8	3
CH_4		g	−74.8	−50.8	186.2	3
			−74.8	−50.8	186.2	6
			−74.8	−50.7	186.3	21
			−74.8	−50.7	186.2	22
			−74.8	−50.7	186.3	31
COS		g	−137.2	−169.2	231.5	3
			−142.1	−169.3	231.6	31

– CERIUM –

Formula	Description	State	ΔH°_f kJ	ΔG°_f kJ	S° J/deg	Source
Ce	Metal	c	0	0	57.7	3
			0	0	75.8	6
			0	0	69.4	11
			0	0	69.5	21
			0	0	72.0	31
Ce^{3+}		aq	−727.2	−713.4	−184.	3
			−697.9	−675.7	−200.8	6

Formula	Description	State	ΔH_f° kJ	ΔG_f° kJ	S° J/deg	Source
			−696.2	−672.0	−205.0	21
			−696.2	−672.0	−205.	31
			−697.9	−675.7	−200.8	119
Ce^{4+}		aq	−577.0	−507.5	−425.5	6
			−537.2	−503.8	−301.	21
			−537.2	−503.8	−301.	31
			−577.0	−507.5	−425.5	119
$Ce(OH)^{3+}$		aq		−785.3		3
			−812.1	−749.4	−170.3	6
			−812.1	−749.4	−170.3	119
$Ce(OH)_2^{2+}$		aq		−1020.9		3
CeO_2	Cerianite	c	−974.9	−916.3	66.1	1
			−1088.6	−1025.3	62.3	6
			−1088.7	−1025.4	62.3	21
			−1088.7	−1024.6	62.3	31
Ce_2O_3		c	−1796.2	−1708.0	148.1	6
			−1796.2	−1707.9	150.6	21
			−1796.2	−1706.2	150.6	31

Formula	Description	State	ΔH_f° kJ	ΔG_f° kJ	S° J/deg	Source
$Ce(OH)_3$		c		-1303.8		1
			-1406.7	-1268.6	110.5	6
CeS		c			78.2	6
			-459.4	-451.5	78.2	31
CeS_2		c	-643.9	-633.9	78.6	1
			-612.1			31
Ce_2S_3		c	-1249.8	-1226.3	131.8	1
					180.3	6
			-1188.			31
$CePO_4$	Monazite	c	-1942.6			4
			-1943.5	-1816.1	115.1	6

- CESIUM -

Formula	Description	State	ΔH_f° kJ	ΔG_f° kJ	S° J/deg	Source
Cs	Metal	c	0	0	82.8	3
			0	0	85.1	6
			0	0	85.2	21
			0	0	85.2	31

Formula	Description	State	ΔH_f° kJ	ΔG_f° kJ	S° J/deg	Source
Cs^+		aq	−247.7	−282.0	133.0	3
			−258.1	−291.8	132.6	6
			−258.0	−283.6	132.8	21
			−258.3	−292.0	133.0	31
					135.8	51
CsO_2		c	−286.2			31
Cs_2O		c	−317.6	−274.5	123.8	1
			−317.6			6
			−346.0	−308.4	146.9	21
			−345.8	−308.1	146.9	31
Cs_2O_2		c	−402.5	−327.2	118.0	1
Cs_2O_3		c	−465.3	−360.2	120.1	1
Cs_2O_4		c	−519.6	−387.0	130.5	1
$Cs(OH)$		c	−406.7	−355.2	77.8	1
			−409.2			6
			−416.7	−370.7	98.7	21
			−417.2			31
Cs_2UO_4		c	−1920.0	−1797.3	219.7	21

Formula	Description	State	ΔH°_f kJ	ΔG°_f kJ	S° J/deg	Source
			-1928.8	-1806.2	219.7	31

- CHLORINE -

Formula	Description	State	ΔH°_f kJ	ΔG°_f kJ	S° J/deg	Source
Cl_2		g	0	0	222.9	3
			0	0	223.0	6
			0	0	223.0	11
			0	0	223.1	21
			0	0	223.1	31
Cl^-		aq	-167.4	-131.2	55.1	3
			-167.2	-131.3	56.5	6
			-167.1	-131.3	56.7	21
			-167.2	-131.2	56.5	31
			-167.3	-131.3		50
			-167.2	-131.2	56.5	119
Cl°_2		aq		6.9		1
			-23.4	6.9	121.3	6
			-23.4	6.9	121.	31

Formula	Description	State	ΔH°_f kJ	ΔG°_f kJ	S° J/deg	Source
			−23.4	6.9	121.3	119
ClO^-		aq		−37.2	41.8	1
			−107.1	−36.8	41.8	6
			−107.1	−36.8	42.	31
			−107.1	−36.8	41.8	119
ClO_2^-		aq	−71.9	11.46	100.8	1
			−66.5	17.2	101.2	6
			−66.5	17.2	101.3	31
			−66.5	17.2	101.2	119
ClO_3^-		aq	−98.3	−2.6	163.2	3
			−104.1	−8.5	163.6	6
			−104.0	−8.0	162.3	31
			−104.1	−8.5	163.6	119
ClO_4^-		aq	−131.4	−10.8	182.0	3
			−129.3	−8.9	183.2	6
			−129.3	−8.5	182.0	31
			−129.3	−8.9	183.2	119
HCl		aq	−167.4	−131.2	55.1	3

Formula	Description	State	ΔH_f° kJ	ΔG_f° kJ	S° J/deg	Source
$HClO^\circ$		aq	−116.4	−80.0	130	1
			−120.9	−79.9	146.	6
				−80.0		12
			−120.9	−79.9	142.	31
			−120.9	−79.9	146.4	119
$HClO_2^\circ$		aq	−57.2	0.3	176	1
			−51.9	28.4	188.3	6
			−51.9	5.9	188.3	31
			−51.9	28.4	188.3	119
$HClO_3^\circ$		aq	−98.3	−2.6	163.2	1
$HClO_4^\circ$		aq	−131.4	−10.3	180.7	1
HCl		g	−92.3	−95.3	186.7	3
			−92.3	−95.3	186.8	6
				−95.3		12
			−92.3	−95.3	186.9	21
			−92.3	−95.3	186.9	31

Formula	Description	State	ΔH°_f kJ	ΔG°_f kJ	S° J/deg	Source
		- CHROMIUM -				
Cr		c	0	0	23.8	3
			0	0	23.8	6
			0	0	23.6	11
			0	0	23.6	21
			0	0	23.8	31
Cr^{2+}		aq	-138.9	-176.1		1
				-164.4		6
			-144.0			21
			-143.5			31
Cr^{3+}		aq	-211.7	-203.9	-316.3	6
			-236.1	-223.2	-215.6	119
Cr^{3+}	$[Cr(6H_2O)]^{3+}$	aq	-256.1	-215.5	-307.5	1
$[Cr(H_2O)_6]^{3+}$		aq	-1999.1			31
CrO_2^-		aq		-518.4		6
				-535.9		12
CrO_3^{3-}		aq		-603.4		12

Formula	Description	State	ΔH_f° kJ	ΔG_f° kJ	S° J/deg	Source
CrO_4^{2-}		aq	−894.3	−736.8	38.5	1
			−875.4	−720.9	46.0	6
			−881.2	−727.8	50.2	31
			−882.2	−729.9	54.0	119
$Cr_2O_7^{2-}$		aq	−1523.0	−1319.6	213.8	1
			−1478.8	−1287.6	254.0	6
			−1490.3	−1301.1	261.9	31
			−1491.9	−1305.4	270.6	119
$Cr(OH)^{2+}$		aq	−445.9	−418.4	−153.1	6
$Cr(OH)^{2+}$	$[Cr(5H_2O)(OH)]^{2+}$	aq	−474.9	−431.0	−68.6	1
$[Cr(H_2O)_5OH]^{2+}$	unspc. aq. soln.	aq	−1939.7			31
$Cr(OH)_2^+$		aq		−623.8		6
				−632.7		12
$HCrO_4^-$		aq	−921.3	−773.6	69.0	1
			−872.5	−758.0	180.3	6
			−878.2	−764.7	184.1	31
$H_2CrO_4^\circ$		aq	−843.9	−752.4	259.8	6
				−777.9		12

Formula	Description	State	ΔH_f° kJ	ΔG_f° kJ	S° J/deg	Source
CrO_2		c	−598.			31
CrO_3		c	−589.5			31
Cr_2O_3	Eskolaite	c	−1128.4	−1046.8	81.2	3
			−1141.0	−1059.4	81.2	6
			−1134.7	−1053.1	81.2	21
			−1139.7	−1058.1	81.2	31
$Cr(OH)_2$		c		−587.8		1
			−659.0	−576.1	81.2	6
$Cr(OH)_3$		c	−1033.9	−900.8	80.3	1
			−975.7	−846.8	95.4	6
	Pptd.		−1064.0			31
			−1014.0	−867.9		119
$Cr(OH)_3$	Hydrous, probably $[Cr(5H_2O)OH](OH)_2$	c	−989.9	−859.8	82.0	1
$[Cr(H_2O)_5OH](OH)_2$		c	−2454.8			31
$Cr(OH)_4$		c		−1014.1		12

Formula	Description	State	ΔH_f° kJ	ΔG_f° kJ	S° J/deg	Source
		- COBALT -				
Co	Metal	c	0	0	28.4	3
			0	0	30.0	6
			0	0	30.0	11
			0	0	30.0	21
			0	0	30.0	31
Co^{2+}		aq	−59.4	−53.6	−113.	1
			−58.6	−56.1	−108.8	6
			−58.2	−54.4	−113.0	21
			−58.2	−54.4	−113.	31
			−58.2	−54.4	−113.	80
			−58.6	−56.1	−108.8	119
Co^{3+}		aq		120.9		1
			25.1	78.2	−316.3	6
			92.0	134.0	−305.0	21
			92.	134.	−305.	31
			25.1	78.2	−317.6	119

Formula	Description	State	ΔH_f° kJ	ΔG_f° kJ	S° J/deg	Source
$Co(OH)^+$		aq		-237.1		33
$Co(OH)_2^o$		aq		-423.1		6
				-421.7		31
				-423.1		119
$Co(OH)_3^-$		aq		-587.8		6
				-587.8		119
$HCoO_2^-$		aq		-347.1		12
				-407.5		31
CoO		c	-231.0	-205.0	43.9	1
			-238.9	-215.0	53.0	6
			-237.9	-214.2	53.0	21
			-237.9	-214.2	53.0	31
CoO_2		c		-216.9		12
Co_3O_4	Cobalt spinel	c	-891.	-774.	102.5	6
				-702.1		12
			-891.2	-772.6	102.5	21
			-891.	-774.	102.5	31
$Co(OH)_2$		c	-541.0	-456.0	82.0	1

Formula	Description	State	ΔH°_f kJ	ΔG°_f kJ	S° J/deg	Source
$Co(OH)_2$	Pink	c	−541.5	−455.2	76.1	6
$Co(OH)_2$	Pink, pptd.	c	−539.7	−454.3	79.	31
$Co(OH)_2$	Transvaalite	c		−459.6		6
$Co(OH)_2$	Pink, pptd., aged	c		−458.1		31
$Co(OH)_2$	Blue, pptd.	c		−450.1		31
$Co(OH)_3$		c	−730.5	−596.6	83.7	1
$Co(OH)_3$	Pptd.	c	−716.7			31
$CoOOH$	Pptd.	c		−386.2		64
CoS α	Pptd.	c	−80.8	−82.8	67.4	1
CoS β		c	−82.8	−84.5	62.3	6
CoS		c	−82.8			31
CoS_2	Cattierite	c	−134.3	−137.2	103.3	6
Co_2S_3		c	−196.6			1
	Pptd.	c	−147.3			31
Co_3S_4	Linnaeite	c	−307.1	−323.4	274.5	6
$CoCO_3$	Sphaerocobaltite	c		−650.9		1
			−716.1	−640.2	88.6	6
				−650.9		12

Formula	Description	State	ΔH°_f kJ	ΔG°_f kJ	S° J/deg	Source
				−648.0		24
			−713.0			31
$CoSO_4$		c	−859.8	−753.5	113.4	1
			−889.4	−783.7	117.6	6
			−888.3	−782.3	118.0	31
$Co(NO_3)_2$		c		−230.5		12
			−420.5			31

- COPPER -

Formula	Description	State	ΔH°_f kJ	ΔG°_f kJ	S° J/deg	Source
Cu	Metal	c	0	0	33.3	3
			0	0	33.3	6
			0	0	33.3	11
			0	0	33.2	21
			0	0	33.2	22
			0	0	33.2	31
Cu^+		aq	51.9	50.2	−26.4	3
			72.1	50.0	42.2	6

Formula	Description	State	ΔH°_f kJ	ΔG°_f kJ	S° J/deg	Source
			71.7	50.0	41.0	21
			71.7	50.0	40.6	31
			72.1	50.0	42.2	119
Cu^{2+}		aq	64.4	65.0	-98.7	3
			65.7	65.3	-95.8	6
			64.8	65.5	-99.6	21
			64.8	65.5	-99.6	31
			65.3	65.7		50
			65.7	65.3	-95.8	119
CuO_2^{2-}		aq		-181.2		3
				-183.6		31
$HCuO_2^-$		aq		-258.5		31
$Cu(OH)^+$		aq		-130.0		6
				-130.0		119
$CuCl^+$		aq		-68.2		5
				-71.4		6
				-68.2		31
				-66.3		104

Formula	Description	State	ΔH_f° kJ	ΔG_f° kJ	S° J/deg	Source
				-71.4		119
$CuCl_2^-$		aq		-240.2		5
				-244.2		6
				-240.1		31
				-240.5		104
				-244.2		119
$CuCl_2$		aq		-197.9		5
				-197.9		31
				-189.2		104
$CuCl_3^{2-}$		aq		-376.6		5
				-376.		31
				-357.9		104
$CuCl_4^{2-}$		aq		-433.5		104
$CuCO_3^o$		aq		-502.1		6
				-501.7		79
				-502.1		119
$Cu(CO_3)_2^{2-}$		aq		-1048.5		6
				-1048.1		79

Formula	Description	State	ΔH°_f kJ	ΔG°_f kJ	S° J/deg	Source
				-1048.5		119
$CuCO_3(OH)_2^{2-}$		aq		-861.9		104
$CuSO_4$		aq		-692.2		5
			-838.5	-691.4	-18.0	6
				-692.2		31
			-838.5	-691.4	-18.0	119
$Cu(NH_3)^{2+}$		aq	-38.9	15.6	12.1	31
				14.6		104
$Cu(NH_3)_2^{2+}$		aq	-142.3	-30.4	111.3	31
				-32.0		104
CuO	Tenorite	c	-155.2	-127.2	43.5	3
			-155.8	-128.0	42.6	6
			-157.3	-129.6	42.6	21
			-155.6	-127.9	42.6	22
			-157.3	-129.7	42.6	31
			-154.4	-124.7	42.6	37
				-128.0		96
				-125.1		97

Formula	Description	State	ΔH_f° kJ	ΔG_f° kJ	S° J/deg	Source
				-125.9		109
Cu_2O	Cuprite	c	-166.7	-146.4	100.8	3
			-170.8	-147.9	92.4	6
			-168.6	-146.0	93.1	21
			-170.8	-148.0	92.4	22
			-168.6	-146.0	93.1	31
$Cu(OH)$		c		-182.0		93
$Cu(OH)_2$		c	-443.9	-356.9	79.	1
			-449.8			31
$CuCl$	Nantokite	c	-134.7	-118.8	91.6	3
			-136.4	-119.2	87.0	6
			-137.2	-119.9	86.2	31
$CuCl_2$		c	-218.8	-176.	112.1	1
			-216.0	-171.8	108.1	6
			-220.1	-175.7	108.1	31
$Cu_4Cl_2(OH)_6$	Atacamite	c		-1341.4		6
				-1339.5		31
			-1657.7	-1341.8	314.6	37

Formula	Description	State	ΔH_f° kJ	ΔG_f° kJ	S° J/deg	Source
$Cu_4Cl_2(OH)_6$	Paratacamite	c		-1338.2		104
				-1341.8		109
CuS	Covellite	c	-48.5	-49.0	66.5	3
			-53.1	-53.6	66.5	6
			-48.6	-49.1	66.6	21
			-52.3	-52.8	66.5	22
			-53.1	-53.6	66.5	31
			-53.3	-53.9		50
Cu_2S	Chalcocite	c	-79.5	-86.2	120.9	3
			-79.5	-86.2	120.9	6
			-80.1	-86.9	120.8	21
			-79.5	-86.3	120.9	22
			-79.5	-86.2	120.9	31
			-80.7	-85.6		50
$CuFeS_2$	Chalcopyrite	c	-177.0	-178.6	118.4	6
			-186.0	-187.9	130.3	22
			-190.4	-190.6		50
Cu_5FeS_4	Bornite	c	-334.4	-362.8	415.4	22

Formula	Description	State	ΔH_f° kJ	ΔG_f° kJ	S° J/deg	Source
			−380.3	−392.9		50
CuSe	Klockmannite	c	−27.6	−33.0	92.9	1
			−39.5	−41.6	82.8	6
			−39.5			31
$CuCO_3$		c	−595.0	−518.0	87.9	3
$Cu_2CO_3(OH)_2$	Malachite	c	−1048.5	−901.3	221.8	6
			−1054.0			21
			−1053.9	−896.2	186.2	22
			−1051.4	−893.6	186.2	31
			−1053.9	−900.4		50
				−905.6		79
				−893.9		95
				−903.7		103
				−905.0		109
$Cu_3(CO_3)_2(OH)_2$	Azurite	c	−1627.6	−1430.9	402.5	6
			−1632.2			21
			−1632.2	−1399.2	280.2	22
			−1632.2	−1315.5	0	31

Formula	Description	State	ΔH°_f kJ	ΔG°_f kJ	S° J/deg	Source
			−1627.6	−1429.7		50
				−1438.2		79
				−1419.4		95
$CuSO_4$	Chalcocyanite	c	−769.8	−661.9	113.4	3
			−769.8	−661.7	112.1	6
			−771.4	−662.3	109.5	21
			−771.4	−661.8	109.	31
Cu_2SO_4		c	−749.8	−652.7	182.4	1
			−751.4			31
$CuSO_4 \cdot H_2O$		c	−1083.6	−917.1	149.8	3
			−1085.8	−918.1	146.0	31
$CuSO_4 \cdot 3H_2O$		c	−1683.1	−1400.0	225.1	3
			−1684.3	−1400.0	221.3	31
$CuSO_4 \cdot 5H_2O$	Chalcanthite	c	−2278.0	−1879.9	305.4	3
			−2278.9	−1879.6		6
			−2279.6	−1879.8	300.4	21
			−2279.6	−1879.7	300.4	31
$Cu_3SO_4(OH)_4$	Antlerite	c		−1446.0		6

Formula	Description	State	ΔH°_f kJ	ΔG°_f kJ	S° J/deg	Source
				−1446.6		31
				−1445.6		79
$Cu_4SO_4(OH)_6$	Brochantite	c		−1817.9		6
				−1818.0		21
				−1817.7		31
			−2198.3	−1817.1	302.5	37
				−1818.4		79
$Cu_4SO_4(OH)_6 \cdot 1.3H_2O$	Langite	c		−1930.9		6
				−2115.0		79
$Cu_4SO_4(OH)_6 \cdot H_2O$	Langite	c	−2485.	−2044.0	335.	31
$Cu_3(PO_4)_2$		c	−2231.4	−2051.6		4
				−2051.3		31
				−2066.6		104
$CuSiO_3 \cdot H_2O$	Dioptase	c	−1359.0	−1207.5	86.6	6
$CuSiO_3 \cdot 2H_2O$	Chrysocolla	am		−1443.9		107

Formula	Description	State	ΔH_f° kJ	ΔG_f° kJ	S° J/deg	Source

- FLUORINE -

Formula	Description	State	ΔH_f° kJ	ΔG_f° kJ	S° J/deg	Source
F_2		g	0	0	203.3	3
			0	0	202.7	6
			0	0	202.8	21
			0	0	202.8	31
F		g	76.6	59.4	158.6	3
			79.0	61.9	158.8	31
F^-		aq	−329.1	−276.5	−9.6	3
			−333.8	−280.0	−14.0	6
			−335.4	−281.7	−13.2	21
			−332.6	−278.8	−13.8	31
			−333.8	−280.0	−14.0	119
HF		aq	−329.1	−294.6	109.	1
			−320.5	−298.1	91.4	6
			−320.1	−296.8	88.7	31
			−320.5	−298.1	91.4	119
HF_2^-		aq	−642.7	−575.3	2.1	1

Formula	Description	State	ΔH°_f kJ	ΔG°_f kJ	S° J/deg	Source
			−660.6	−581.5	67.8	6
			−649.9	−578.1	92.5	31
			−660.6	−581.5	67.8	119
F_2O		g	23.0	40.6	246.6	1
			24.7	41.9	247.4	31
HF		g	−268.6	−270.7	173.5	3
			−271.1	−273.2	173.7	6
			−273.3	−275.4	173.8	21
			−271.1	−273.2	173.8	31
HF		1	−303.3			6
			−299.8			31

- GERMANIUM -

Formula	Description	State	ΔH°_f kJ	ΔG°_f kJ	S° J/deg	Source
Ge		c	0	0	31.1	6
			0	0	42.2	11
			0	0	31.1	21
			0	0	31.1	31

Formula	Description	State	ΔH_f° kJ	ΔG_f° kJ	S° J/deg	Source
Ge^{2+}		aq		−36.8		6
				0		12
Ge^{4+}		aq		−27.6		6
GeO_3^{2-}		aq		−660.6		12
$HGeO_3^-$		aq		−733.0		12
$H_2GeO_3^o$		aq		−781.6		12
	unspc. aq. soln.		−818.9			31
$HGeO_2^-$		aq		−385.3		12
GeO	Brown	c	−237.2	−212.1	50.2	6
			−261.9	−237.2	50.	31
GeO	Hydrated, a	c		−292.5		12
GeO	Yellow	c		−207.1		6
				−207.1		31
GeO	Hydrated, b	c		−262.3		12
GeO_2	Hexagonal	c	−554.7	−500.8	55.3	6
	Quartz type		−551.0	−497.1	55.3	21
			−551.0	−497.0	55.3	31
GeO_2	Pptd., hexagonal	c		−552.		12

Formula	Description	State	ΔH_f° kJ	ΔG_f° kJ	S° J/deg	Source
GeO_2		c	−580.2	−521.6	39.7	6
GeO_2	Tetragonal	c		−569.4		12
GeO_2		am	−540.1			6
			−526.4	−475.2	64.5	21
			−537.2			31

- GOLD -

Formula	Description	State	ΔH_f° kJ	ΔG_f° kJ	S° J/deg	Source
Au	Metal	c	0	0	47.7	3
			0	0	47.6	6
			0	0	47.3	11
			0	0	47.5	21
			0	0	47.4	22
			0	0	47.4	31
Au^+		aq		163.2		1
			222.2	178.6	127.6	6
			222.2	178.6	127.6	119
Au^{3+}		aq		433.5		1

Formula	Description	State	ΔH°_f kJ	ΔG°_f kJ	S° J/deg	Source
AuO_3^{3-}		aq		−24.3		3
				−51.8		31
$Au(OH)_3^o$		aq		−317.6		6
				−283.4		31
				−317.6		119
$HAuO_3^{2-}$		aq		−115.5		3
				−142.2		31
$H_2AuO_3^-$		aq		−191.6		3
				−218.3		31
$H_3AuO_3^o$		aq		−258.6		3
$AuCl_2^-$		aq	−174.5	−151.2	257.3	6
				−151.1		31
			−174.5	−151.2	257.3	119
$AuCl_4^-$		aq	−324.7	−235.0	257.7	6
			−322.2	−235.1	266.9	31
			−324.7	−235.0	257.7	119
AuO_2		c		200.8		12
Au_2O_3		c	80.8	163.2	126.	3

Formula	Description	State	ΔH°_f kJ	ΔG°_f kJ	S° J/deg	Source
			-2.6	77.2	134.3	6
$Au(OH)_3$		c	-418.4	-290.0	121.	3
			-478.2	-348.9	116.7	6
			-424.7	-316.9	189.5	31

- HYDROGEN -

Formula	Description	State	ΔH°_f kJ	ΔG°_f kJ	S° J/deg	Source
H_2		g	0	0	130.6	3
			0	0	130.6	6
			0	0	133.5	7
			0	0	130.5	11
			0	0	130.7	21
			0	0	130.6	22
			0	0	130.7	31
H^+		aq	0	0	0	6
			0	0	0	21
			0	0	0	31
H_2		aq	-4.2	17.6	57.7	6

Formula	Description	State	ΔH°_f kJ	ΔG°_f kJ	S° J/deg	Source
			−4.2	17.6	57.7	31
			−4.2	17.6	57.7	119

— IODINE —

Formula	Description	State	ΔH°_f kJ	ΔG°_f kJ	S° J/deg	Source
I_2		c	0	0	116.7	3
			0	0	116.1	6
			0	0	116.2	21
			0	0	116.1	31
I_2		g	62.2	19.4	260.6	3
			62.4	19.4	260.6	6
			62.4	19.3	260.7	21
			62.4	19.3	260.7	31
I^-		aq	−55.9	−51.7	109.4	3
			−57.7	−51.6	102.9	6
			−56.9	−51.9	106.7	21
			−55.2	−51.6	111.3	31
			−57.7	−51.6	102.9	119

Formula	Description	State	ΔH_f° kJ	ΔG_f° kJ	S° J/deg	Source
I_2		aq	20.9	16.4		1
			21.8	16.4	134.7	6
			22.6	16.4	137.2	31
			21.8	16.4	134.7	119
I_3^-		aq	−51.9	−51.5	173.6	3
			−52.5	−51.6	236.4	6
			−51.5	−51.4	239.3	31
			−52.5	−51.6	236.4	119
I_5^-		aq		−28.9		12
IO^-		aq	−142.	−35.6		1
			−107.5	−38.5	−5.4	31
IO_3^-		aq	−230.1	−135.6	115.9	3
			−219.7	−125.9	117.2	6
			−221.3	−128.0	118.4	31
			−219.7	−125.9	117.2	119
IO_4^-		aq		−53.1		12
			−151.5	−58.5	222.	31
IO_5^{3-}		aq		−180.4		12

Formula	Description	State	ΔH°_f kJ	ΔG°_f kJ	S° J/deg	Source
HIO		aq	−159.	−98.3		1
			−138.1	−99.1	95.4	31
HIO_3		aq		−139.5		12
			−211.3	−132.6	166.9	31
HIO_4		aq		−62.8		12
HIO_5^{2-}		aq		−243.1		12
H_2IO^+		aq		−106.3		12
				−106.7		31
$H_4IO_6^-$		aq		−518.3		12
			−759.4			31
ICl		aq		−16.7		1
				−17.1		31
ICl		c	−33.6	−13.6	102.5	1
			−35.1			31
ICl_3		c	−88.3	−22.4	172.0	3
			−89.5	−22.3	167.4	31

Formula	Description	State	ΔH_f° kJ	ΔG_f° kJ	S° J/deg	Source
	- IRON -					
Fe	Metal	c	0	0	27.2	3
			0	0	27.2	6
			0	0	27.2	7
			0	0	27.2	11
			0	0	27.3	21
			0	0	27.3	31
Fe^{2+}		aq	−87.9	−84.9	−113.4	3
			−92.6	−92.2	−104.6	6
			−92.6	−90.1	−111.7	7
			−89.1	−78.9	−138.0	21
			−89.1	−78.9	−137.7	31
			−89.1	−91.2		50
			−92.0	−91.3	−105.8	80
			−92.6	−92.2	−104.6	119
Fe^{3+}		aq	−47.7	−10.5	−293.3	3
			−50.8	−17.9	−279.1	6

Formula	Description	State	ΔH°_f kJ	ΔG°_f kJ	S° J/deg	Source
			−50.8	−15.7	−286.2	7
			−48.5	−4.6	−316.0	21
			−48.5	−4.7	−315.9	31
			−46.4	−16.9		50
			−49.4	−17.0	−277.4	80
			−50.8	−17.9	−279.1	119
FeO_4^{2-}		aq	−479.5			6
				−467.3		12
			−479.5			119
FeO_2H^-		aq		−379.2		12
				−377.7		31
$Fe(OH)^+$		aq	−328.2	−291.0	4.6	6
			−328.2	−280.0	−32.2	7
			−324.7	−277.4	−29.	31
				−268.6		60
$Fe(OH)^{2+}$		aq	−292.9	−242.7	−104.2	6
			−291.8	−240.4	−107.9	7
			−290.8	−229.4	−142.	31

Formula	Description	State	ΔH_f° kJ	ΔG_f° kJ	S° J/deg	Source
$Fe(OH)_2^+$		aq		−444.3		3
				−453.6		6
			−551.3	−457.2	−18.0	7
				−438.0		31
$Fe(OH)_2^\circ$		aq		−459.2		6
				−457.1		7
$Fe(OH)_3^-$		aq		−621.2		6
				−619.2		7
				−614.9		31
$Fe(OH)_4^{2-}$		aq		−776.2		6
				−774.0		7
				−769.7		31
$Fe(OH)_3^\circ$		aq		−677.4		6
				−670.7		7
				−659.3		31
$Fe(OH)_4^-$		aq		−843.9		6
				−841.8		7
$Fe_2(OH)_2^{4+}$		aq	−612.1	−467.3	−356.	31

Formula	Description	State	ΔH_f° kJ	ΔG_f° kJ	S° J/deg	Source
$FeCl^{2+}$		aq	−179.5	−150.2	−92.	3
			−194.5	−157.6	−115.9	6
			−180.3	−143.9	−113.	31
$Fe_{0.95}O$	Wustite	c	−266.5	−244.3	54.0	3
			−266.3	−245.1	57.5	6
			−266.9	−245.8	57.5	11
			−266.3	−245.2	57.6	21
			−266.3	−245.1	57.5	31
			−265.4	−244.6	58.5	108
FeO	Wustite	c	−265.4	−244.3	58.9	7
FeO	Stoichiometric	c	−272.7	−252.1		11
			−272.0	−251.2	59.8	21
			−272.0	−251.4	60.8	22
			−272.0			31
Fe_2O_3	Hematite	c	−822.2	−741.0	90.0	3
			−824.2	−742.4	87.4	6
			−821.4	−739.6	87.4	7
				−743.5		9

Formula	Description	State	ΔH°_f kJ	ΔG°_f kJ	S° J/deg	Source
			−823.4	−741.6	87.4	11
			−824.6	−742.7	87.4	21
			−827.3	−745.4	87.6	22
			−824.2	−742.2	87.4	31
			−827.1	−745.3	87.5	108
Fe_3O_4	Magnetite	c	−1117.1	−1014.2	146.4	3
			−1118.4	−1015.4	146.4	6
			−1112.4	−1010.8	150.7	7
			−1118.8	−1017.3	150.7	11
			−1115.7	−1012.6	146.1	21
			−1118.4	−1015.4	146.4	31
			−1117.3	−1014.1	146.0	108
$Fe(OH)_2$	Amakinite	c	−568.2	−483.5	79.	3
			−573.2	−493.0	92.5	6
				−491.0		7
	Pptd.		−569.0	−486.5	88.	31
$Fe(OH)_3$		c	−824.2	−694.5	96.	1
			−844.3	−714.6	96.	6

Formula	Description	State	ΔH°_f kJ	ΔG°_f kJ	S° J/deg	Source
				-718.6		7
				-694.5		19
	Pptd.		-823.0	-696.5	106.7	31
FeO(OH)	Goethite	c	-559.0	-490.4	67.4	6
			-560.0	-489.7	61.5	7
			-559.3	-488.6	60.4	21
			-559.0			31
			-559.3	-490.1		50
FeO(OH)	Lepidocrocite	c		-471.4		7
$FeCl_2$	Lawrencite	c	-341.8	-303.0	117.9	6
			-341.6	-302.2	118.0	21
			-341.8	-302.3	118.0	31
$FeCl_3$	Molysite	c	-405.0	-336.4	130.1	1
			-399.5	-331.8	134.7	6
			-399.2	-333.8	142.3	21
			-399.5	-334.0	142.3	31
FeS	α	c	-95.1	-97.6	67.4	3
	Troilite, α		-100.1	-100.5	60.3	6

Formula	Description	State	ΔH_f° kJ	ΔG_f° kJ	S° J/deg	Source
FeS	Troilite	c	−101.0	−101.3	60.3	7
			−101.0	−101.3	60.3	21
FeS	Pyrrhotite	c	−100.4	−100.8	60.3	22
	Iron-rich pyrrhotite	c	−100.0	−100.4	60.3	31
$Fe_{0.877}S$	Pyrrhotite	c			60.8	6
					60.8	21
			−105.4	−107.1		50
	Sulfur-rich pyrrhotite	c	−92.0	−93.6	60.7	31
FeS	Mackinawite, synthetic	c		−93.3		7
FeS_2	Pyrite	c	−174.0	−162.8	52.9	6
			−171.5	−160.2	52.9	7
			−171.5	−160.2	52.9	21
			−171.5	−160.2	52.9	22
			−178.2	−166.9	52.9	31
			−173.6	−162.3		50
				−150.6		81
FeS_2	Marcasite	c	−150.6			6
			−169.4	−158.4	53.9	21

Formula	Description	State	ΔH°_f kJ	ΔG°_f kJ	S° J/deg	Source
			-154.8			31
FeAsS	Arsenopyrite	c	-105.4	-109.6	108.4	6
			-42.	-50.	121.	31
FeSe	Pptd.	c	-69.0	-58.2		3
			-75.3	-75.7	70.6	6
			-75.3			31
$FeSe_2$	Ferroselite	c			86.8	6
					86.9	21
					86.8	31
$FeTe_2$	Frohbergite	c			100.2	6
					100.2	21
					100.2	31
Fe_3C	Cohenite	c	24.2	18.1	107.5	6
			24.9	19.9	104.4	21
Fe_3C α	Cementite	c	25.1	20.1	104.6	31
$FeCO_3$	Siderite	c	-747.7	-673.9	92.9	3
			-753.1	-680.3	96.1	6
			-752.2	-679.5	96.1	7

Formula	Description	State	ΔH_f° kJ	ΔG_f° kJ	S° J/deg	Source
			−737.0	−666.7	105.0	21
			−749.6	−679.4	105.0	22
			−740.6	−666.7	92.9	31
			−737.0	−679.4		50
			−749.6	−679.5		80
$(Fe_{0.956}Mn_{0.042}Mg_{0.002})CO_3$	Siderite, natural	c			95.5	106
$FeSO_4$		c	−922.6	−829.7	115.5	1
			−928.4	−820.5	107.5	6
			−928.4	−820.8	107.5	31
$Fe_2(SO_4)_3$		c	−2581.9	−2246.8	259.0	6
			−2576.9	−2250.0	282.8	21
			−2581.5			31
$FeSO_4 \cdot H_2O$	Szomolnokite	c	−1244.0			6
			−1243.9			21
			−1243.7			31
$FeSO_4 \cdot 7H_2O$	Melanterite	c	−3014.5	−2510.2	409.2	6
			−3014.4	−2509.6	409.2	21
			−3014.6	−2509.9	409.2	31

Formula	Description	State	ΔH°_f kJ	ΔG°_f kJ	S° J/deg	Source
$FePO_4$	Heterosite	c	−1253.5	−1138.	93.7	1
				−1148.4		4
			−1297.4	−1184.9	100.8	6
			−1297.5			31
$FePO_4 \cdot 2H_2O$	Strengite	c	−1888.1	−1657.5	171.3	6
			−1888.2	−1662.9	171.2	21
			−1888.2	−1657.5	171.2	31
$Fe_3(PO_4)_2 \cdot 8H_2O$	Vivianite	c		−4377.2		4
$FeMoO_4$		c	−1077.4	−982.4	139.7	1
			−1074.4	−974.3	129.3	6
			−1075.	−975.	129.3	31
$FeWO_4$	Ferberite	c	−1146.8	−1047.7	148.1	1
			−1187.8	−1087.0	131.8	6
			−1154.8	−1053.9	131.8	21
			−1155.	−1054.	131.8	31
$FeAl_2O_4$	Hercynite	c	−1975.1	−1859.5	106.3	6
			−1981.1	−1864.8	106.3	11
			−1966.5	−1850.8	106.3	21

Formula	Description	State	ΔH°_f kJ	ΔG°_f kJ	S° J/deg	Source
			−1995.3	−1879.8	106.3	31
$FeCr_2O_4$	Chromite	c	−1453.6	−1352.6	146.0	6
			−1460.6	−1359.8	146.0	11
					146.0	21
			−1461.5	−1360.6	146.4	119
$FeSiO_3$		c	−1196.2			6
			−1205.			31
$FeSiO_3$	Ferrosilite	c	−1155.	−1075.	−87.4	1
			−1205.8	−1123.2	86.2	11
			−1195.0	−1117.8	94.6	22
			−1195.2	−1119.6	90.5	80
			−1194.3	−1117.4	96.0	108
$FeSiO_3$	Clinoferrosilite	c	−1196.2	−1118.5	92.8	7
			−1197.1	−1119.7	93.9	11
$FeSiO_3$	Orthoferrosilite	c	−1192.4	−1114.6	92.9	11
Fe_2SiO_4	Fayalite	c	−1438.0	−1338.0	148.1	3
			−1480.2	−1379.4	145.2	6

Formula	Description	State	ΔH°_f kJ	ΔG°_f kJ	S° J/deg	Source
			-1484.6	-1383.8	145.2	7
			-1478.8	-1377.9	145.2	11
			-1479.4	-1379.4	148.3	21
			-1481.6	-1381.6	148.3	22
			-1479.9	-1379.0	145.2	31
			-1479.4	-1379.4	148.3	47
			-1479.2	-1380.2	151.5	108
Fe_2SiO_4	Fayalitic Spinel	c	-1472.3	-1368.6	135.6	11
$Fe_3Si_2O_5(OH)_4$	Greenalite	c		-3012.		8
$Fe_3Si_4O_{10}(OH)_2$	Minnesotaite	c	-4822.5	-4476.5	356.5	7
$Fe_7Si_8O_{22}(OH)_2$	Grunerite	c	-9598.1	-8957.9	782.4	7
$Fe_3Al_2Si_3O_{12}$	Almandine	c	-5301.8	-4969.8	311.3	11
$Fe_2Al_4Si_5O_{18}$	Ferrocordierite	c	-8450.5	-7960.9	469.6	11
$FeAl_2SiO_5(OH)_2$	Chloritoid	c	-3191.3	-2966.8	197.8	11
$Fe_2Al_9Si_4O_{23}(OH)$	Staurolite	c	-12058.9	-11191.2	589.1	11
					509.8	67
$(H_3Al_{1.15}Fe^{2+}_{0.6})(Fe^{2+}_{2.07}Fe^{3+}_{0.54}Ti_{0.08}$ $Mn_{0.02}Al_{1.19})(Mg_{0.44}Al_{15.26})Si_8O_{48}$	Staurolite	c			1101.0	67

Formula	Description	State	ΔH_f° kJ	ΔG_f° kJ	S° J/deg	Source
$Fe_4Al_4Si_2O_{10}(OH)_8$	Daphnite	c	−7630.4	−7022.5	587.8	11
$Fe_5Al_2Si_3O_{10}(OH)_8$	Fe-clinochlore	c		−7536.3		24
$FeMgSi_2O_6$	Hypersthene	c	−2755.6	−2592.8		24
$FeTiO_3$	Ilmenite	c	−1235.5	−1158.2	105.8	6
			−1236.6	−1159.2	105.9	21
			−1236.6	−1159.2	105.8	119
Fe_2TiO_4	Ulvospinel	c			169.0	6
Fe_2TiO_4	Titanomagnetite	c			168.9	21
Fe_2TiO_4		c			163.2	119
Fe_2TiO_5	Pseudobrookite	c			156.5	6
					156.5	21
					156.5	119

– LANTHANUM –

Formula	Description	State	ΔH_f° kJ	ΔG_f° kJ	S° J/deg	Source
La	Metal	c	0	0	57.3	3
			0	0	56.6	6
			0	0	56.9	21

Formula	Description	State	ΔH°_f kJ	ΔG°_f kJ	S° J/deg	Source
La^{3+}			0	0	56.9	31
		aq	−737.2	−730.1	−163.2	1
			−707.1	−686.2	−209.2	6
			−707.1	−683.7	−217.6	31
			−707.1	−686.2	−209.2	119
La$_2$O$_3$		c	−1916.3	−1786.1	121.8	1
			−1793.3	−1705.4	127.3	6
			−1793.7	−1706.0	127.3	21
			−1793.7	−1705.8	127.3	31
La(OH)$_3$		c	−1443.5	−1310.4	105.	1
			−1410.0			31
La(OH)$_3$		am	−1410.0	−1286.2	144.8	6
LaS		c	−456.	−451.5	73.2	31
LaS$_2$		c	−655.6	−647.3	78.6	1
La$_2$S$_3$		c	−1283.6	−1260.2	131.8	1
			−1209.			31

Formula	Description	State	ΔH_f° kJ	ΔG_f° kJ	S° J/deg	Source
			– LEAD –			
Pb	Metal	c	0	0	64.9	3
			0	0	64.8	6
			0	0	65.1	21
			0	0	64.8	31
Pb^{2+}		aq	−1.6	−24.3	21.3	3
			−1.4	−24.4	11.3	6
			−1.7	−24.4	10.0	21
			−1.7	−24.4	10.5	31
			−1.7	−24.4		50
			−1.4	−24.4	11.3	119
Pb^{4+}		aq		302.5		1
PbO_3^{2-}		aq		−277.6		12
PbO_4^{4-}		aq		−282.1		12
$Pb(OH)^+$		aq		−226.4		6
				−226.3		31
				−224.1		32

Formula	Description	State	ΔH_f° kJ	ΔG_f° kJ	S° J/deg	Source
				−220.1		34
				−226.4		119
$Pb(OH)_2^o$		aq		−400.8		6
				−408.4		32
				−402.1		34
				−400.8		119
$Pb(OH)_3^-$		aq		−575.7		6
				−575.6		31
				−575.7		119
$HPbO_2^-$		aq		−338.9		1
				−338.4		31
$PbCl^+$		aq		−164.9		7
				−164.8		31
$PbCl_2$		aq		−300.8		7
				−297.2		31
$PbCl_3^-$		aq		−426.3		31
PbO	Litharge, red	c	−219.2	−189.3	67.8	3
			−219.0	−188.9	66.5	6

Formula	Description	State	ΔH°_f kJ	ΔG°_f kJ	S° J/deg	Source
			−219.4	−189.2	66.3	21
			−219.0	−188.9	66.5	31
PbO	Massicot, yellow	c	−217.9	−188.5	69.4	3
			−217.3	−187.9	68.7	6
			−218.1	−188.6	68.7	21
			−217.3	−187.9	68.7	31
				−187.9		101
PbO_2	Plattnerite	c	−276.6	−219.0	76.6	3
			−276.6	−217.6	71.8	6
			−274.5	−215.3	71.8	21
			−277.4	−217.3	68.6	31
				−217.4		101
Pb_2O_3		c			151.9	6
				−411.8		12
					151.9	31
Pb_3O_4	Minium	c	−734.7	−617.6	211.3	3
			−718.7	−601.4	212.0	21
			−718.4	−601.2	211.3	31

Formula	Description	State	ΔH_f° kJ	ΔG_f° kJ	S° J/deg	Source
				−601.2		101
Pb(OH)$_2$		c	−514.6	−420.9	88.	3
				−452.3		6
				−452.2		31
PbCl$_2$	Cotunnite	c	−359.2	−314.0	136.4	3
			−359.4	−314.1	136.0	6
			−359.4	−314.0	136.0	21
			−359.4	−314.1	136.0	31
			−359.4	−314.2		102
PbClOH	Laurionite	c		−480.3		6
				−391.2		31
			−460.5	−391.2		102
PbF$_2$		c	−663.2	−619.6	121.	3
			−678.6	−627.6	96.2	6
			−664.0	−617.1	110.5	31
PbBr$_2$		c	−277.0	−260.4	161.5	3
			−278.6	−261.9	161.5	6
			−278.7	−261.9	161.5	31

Lead

Formula	Description	State	ΔH_f° kJ	ΔG_f° kJ	S° J/deg	Source
PbI_2		c	−175.1	−173.8	177.0	3
			−178.1	−176.4	175.2	6
			−175.5	−173.6	174.8	31
PbS	Galena	c	−94.3	−92.7	91.2	3
			−100.4	−98.7	91.2	6
			−97.7	−96.1	91.4	21
			−98.3	−96.7	91.2	22
			−100.4	−98.7	91.2	31
				−98.7		101
$PbSe$	Clausthalite	c	−75.3	−64.4	112.5	1
			−102.9	−101.7	102.5	6
			−102.9	−101.6	102.5	21
			−102.9	−101.7	102.5	31
$PbTe$	Altaite	c	−73.2	−75.7	115.5	1
			−68.5	−69.4	117.6	6
			−70.7	−69.4	110.0	21
			−70.7	−69.5	110.0	31
PbH_2		g		290.8		12

Formula	Description	State	ΔH°_f kJ	ΔG°_f kJ	S° J/deg	Source
$PbCO_3$	Cerussite	c	−700.0	−626.3	131.0	3
			−702.7	−629.0	131.0	6
				−626.3		12
			−699.2	−625.3	131.0	21
			−702.9	−629.1	131.0	22
			−699.1	−625.5	131.0	31
				−628.8		32
				−625.5		101
$PbO \cdot PbCO_3$		c	−920.5	−818.4	202.9	3
			−918.4	−816.7	204.2	31
$2PbO \cdot PbCO_3$		c	−1142.	−1012.	272.	3
$Pb_3(CO_3)_2(OH)_2$	Hydrocerussite	c		−1711.0		6
				−1714.2		32
			−1914.2			50
				−1699		82
				−1700.4		101
$PbSO_4$	Anglesite	c	−918.4	−811.2	147.3	3
			−919.3	−812.5	148.6	6

Formula	Description	State	ΔH°_f kJ	ΔG°_f kJ	S° J/deg	Source
			−919.9	−813.0	148.6	21
			−919.9	−813.2	148.6	22
			−919.9	−813.1	148.6	31
$PbSO_4 \cdot PbO$	Lanarkite	c	−1182.0	−1083.2	203.8	1
			−1171.5	−1032.2	206.7	6
			−1171.5	−1032.1	206.7	31
PbS_2O_3		c	−628.0	−560.6	148.1	1
			−674.0			31
$PbSeO_3$	Molybdomenite	c	−534.3			6
			−537.6			31
$PbSeO_4$		c	−619.	−510.	155.	1
			−616.7	−504.8	142.2	6
			−609.2	−504.9	167.8	31
$Pb_3(PO_4)_2$		c	−2595.3	−2432.6	353.3	3
				−2364.0		4
					353.1	6
					353.1	31
$PbHPO_3$		c	−981.1	−871.5	133.5	1

Formula	Description	State	ΔH°_f kJ	ΔG°_f kJ	S° J/deg	Source
			-979.1			31
$Pb_5(PO_4)_3Cl$	Pyromorphite	c		-3791.5		4
				-3506.6		6
$PbMoO_4$	Wulfenite	c	-1112.1	-969.4	161.1	1
			-1049.0	-948.5	166.1	6
			-1051.9	-951.1	166.1	21
$PbWO_4$	Stolzite	c	-1121.7	-1020.5	168.2	6
					168.2	21
$PbCrO_4$	Crocoite	c	-942.2	-851.9	152.7	1
			-915.0	-816.9	169.4	6
$PbSiO_3$	Alamosite	c	-1082.8	-1000.0	113.	3
			-1145.7	-1062.2	109.6	6
			-1134.1	-1051.0	113.0	11
			-1145.7	-1062.1	109.6	31
			-1145.4	-1061.3	109.6	47
			-1150.5	-1066.5	109.6	119
Pb_2SiO_4		c	-1308.3	-1195.4	179.9	3
			-1363.1	-1252.7	186.6	6

Formula	Description	State	ΔH°_f kJ	ΔG°_f kJ	S° J/deg	Source
			-1359.4	-1246.4	183.2	11
			-1363.1	-1252.6	186.6	31
			-1369.8	-1259.0	186.6	119

- LITHIUM -

Formula	Description	State	ΔH°_f kJ	ΔG°_f kJ	S° J/deg	Source
Li	Metal	c	0	0	28.0	3
			0	0	29.1	6
			0	0	29.1	21
			0	0	29.1	31
Li^+		aq	-278.5	-293.8	14.2	3
			-278.5	-292.6	11.3	6
			-278.4	-292.6	11.3	21
			-278.5	-293.3	13.4	31
			-278.5	-292.6	11.3	119
LiOH		aq	-508.4	-451.1	3.8	3
			-508.4	-451.8	7.1	31
Li_2SO_4		aq	-1464.4	-1329.6	45.6	3

Formula	Description	State	ΔH_f° kJ	ΔG_f° kJ	S° J/deg	Source
$LiNO_3$		aq	−485.0	−404.3	160.7	3
	dissociated, see preface		−485.8	−404.5	160.2	31
	associated, " "			−406.6		31
Li_2O		c	−595.8	−560.2	37.9	1
			−598.5	−561.9	37.9	6
			−598.7	−562.0	37.6	21
			−597.9	−561.2	37.6	31
			−597.5	−561.5	38.1	119
Li_2O_2		c	−634.7	−565.	33.	1
			−634.3			31
LiCl		c	−487.2	−443.9	50.	3
			−484.9	−439.1	42.8	6
			−484.9	−438.9	42.8	21
			−484.9	−439.0	42.8	31
			−487.4	−443.1	42.7	119
LiCl		c	−408.8	−383.7	55.2	1
			−408.6	−384.4	59.3	6
			−408.6	−384.4	59.3	31

Formula	Description	State	ΔH°_f kJ	ΔG°_f kJ	S° J/deg	Source
LiH		c	-90.4	-70.0	24.7	3
			-90.5	-68.4	20.0	31
LiH		g	128.4	105.4	170.6	3
			139.2	116.5	170.9	31
Li_2CO_3		c	-1215.6	-1132.4	90.4	3
			-1212.5	-1128.6	90.4	6
			-1215.9	-1132.1	90.4	31
Li_2SO_4 β		c	-1434.4	-1324.6	113.	1
			-1435.9	-1320.8	114.0	6
			-1436.5	-1321.7	115.1	31
Li_2SO_4		c	-1434.4	-1319.6	114.2	119
$LiNO_3$		c	-482.3	-389.5	105.4	1
			-483.3			6
			-483.1	-381.1	90.0	31
$LiAlO_2$		c	-1190.0	-1127.6	53.3	6
			-1188.7	-1126.3	53.4	21
			-1193.3	-1131.3	53.3	31
			-1189.6			119

Formula	Description	State	ΔH°_f kJ	ΔG°_f kJ	S° J/deg	Source
Li_2SiO_3		c		-1558.7		24
			-1648.1	-1557.2	79.8	31
			-1631.8	-1542.6	83.7	119
$Li_2Si_2O_5$		c	-2559.8	-2414.5	122.2	31
			-2520.9	-2378.2	125.5	119
$LiAlSiO_4$	Eucryptite	c	-2114.0	-2000.0	103.8	6
			-2123.3	-2009.2	103.8	21
			-2124.2	-2010.3	103.8	31
			-2123.3	-2009.2	103.8	47
$LiAlSi_2O_6$ α	Spodumene	c	-3045.1	-2872.0	129.3	6
			-3053.5	-2880.2	129.3	21
			-3054.7	-2881.5	129.3	31
			-3053.5	-2880.2	129.3	47
$LiAlSi_2O_6$ β	Spodumene	c	-3017.1	-2851.3	154.4	6
			-3025.3	-2859.5	154.4	21
			-3026.7	-2860.9	154.4	31
			-3025.3	-2859.5	154.4	47

Formula	Description	State	ΔH_f° kJ	ΔG_f° kJ	S° J/deg	Source
		- MAGNESIUM -				
Mg	Metal	c	0	0	32.5	3
			0	0	32.7	6
			0	0	32.6	7
			0	0	32.7	11
			0	0	32.7	21
			0	0	32.7	31
Mg^{2+}		aq	-462.0	-456.0	-118.0	3
			-461.7	-455.3	-119.7	6
			-461.7	-455.2	-119.7	7
			-466.8	-454.8	-138.0	21
			-466.8	-454.8	-138.1	31
$Mg(OH)^+$		aq		-628.0		2
				-627.6		6
				-638.1		7
				-626.7		31
				-624.5		72

Formula	Description	State	ΔH°_f kJ	ΔG°_f kJ	S° J/deg	Source
				-638.1		119
$MgCO_3^\circ$		aq		-1003.5		2
				-1002.5		6
				-999.3		72
				-1002.5		119
$MgHCO_3^+$		aq		-1049.7		2
				-1050.6		7
				-1047.2		31
$MgSO_4^\circ$		aq		-1211.5		2
			-1368.9	-1211.8	-52.3	6
			-1356.0	-1212.2	-7.1	31
			-1368.9	-1249.5	-52.3	119
MgO	Finely divided	c	-598.1	-566.1	27.9	3
	Microcrystalline		-598.0	-566.0	27.9	31
MgO	Periclase	c	-601.8	-569.6	26.8	3
			-601.7	-569.4	27.0	6
			-601.6	-569.4	26.9	7
			-601.6	-569.4	26.9	11

Formula	Description	State	ΔH°_f kJ	ΔG°_f kJ	S° J/deg	Source
			−601.5	−569.2	26.9	21
			−601.6	−569.4	26.9	22
			−601.7	−569.4	26.9	31
			−601.2	−569.0	27.0	108
$Mg(OH)_2$	Brucite	c	−924.7	−833.7	63.1	3
			−924.5	−833.6	63.1	6
			−927.9	−837.0	63.1	7
			−927.9	−836.9	63.1	11
			−924.5	−833.5	63.2	21
			−926.3	−835.3	63.1	22
			−924.5	−833.5	63.2	31
				−831.4		72
			−925.3	−834.3	63.2	108
$MgCl_2$	Chloromagnesite	c	−641.8	−592.3	89.5	3
			−641.3	−591.8	89.5	6
			−641.3	−591.8	89.6	21
			−641.3	−591.8	89.6	31
MgOHCl		c	−800.4	−732.2	82.8	3

Formula	Description	State	ΔH°_f kJ	ΔG°_f kJ	S° J/deg	Source
			-799.6	-731.7	83.7	31
MgF_2	Sellaite	c	-1124.2	-1071.1	57.2	6
			-1124.2	-1071.1	57.2	21
			-1123.4	-1070.2	57.2	31
$MgBr_2$		c	-517.6	-499.2	123.0	1
			-524.3	-503.8	117.2	31
MgS		c	-347.	-349.8	52.7	1
			-353.1			6
			-346.0	-341.8	50.3	31
MgH		g	172.	142.	199.2	3
$MgCO_3$	Magnesite	c	-1113.	-1029.	65.7	3
			-1095.8	-1012.3	65.7	6
			-1113.1	-1029.7	65.7	7
			-1113.2	-1029.6	65.8	11
			-1113.3	-1029.5	65.1	21
			-1111.4	-1027.8	65.7	22
			-1095.8	-1012.1	65.7	31
				-1027.3		72

Formula	Description	State	ΔH_f° kJ	ΔG_f° kJ	S° J/deg	Source
$MgCO_3 \cdot 3H_2O$	Nesquehonite	c	−1985.7	−1726.6	159.0	6
				−1726.6		7
			−1977.3	−1723.7	195.6	21
			−1977.2	−1724.0	195.6	22
				−1726.1		31
$MgCO_3 \cdot 5H_2O$	Lansfordite	c		−2200.2		6
				−2201.0		7
				−2199.2		31
$Mg_2(OH)_2(CO_3) \cdot 3H_2O$	Artinite	c	−2920.6	−2568.3	232.9	21
			−2920.6	−2568.6	232.9	22
$Mg_5(CO_3)_4(OH)_2 \cdot 4H_2O$	Hydromagnesite	c	−6514.9	−5864.2	503.7	21
			−6514.9	−5864.6	541.3	22
$Mg_4(CO_3)_3(OH)_2 \cdot 3H_2O$		c		−4603.3		31
				−4637.1		65
$MgSO_4$		c	−1278.2	−1173.6	91.6	3
			−1280.0	−1165.8	91.6	6
			−1284.9	−1170.6	91.6	31
$MgSO_4 \cdot 7H_2O$	Epsomite	c	−3387.8	−2870.2	370.3	6

Formula	Description	State	ΔH_f° kJ	ΔG_f° kJ	S° J/deg	Source
			−3388.7	−2871.2	372.0	21
			−3388.7	−2871.5	372.	31
				−2869.9		72
$Mg_3(PO_4)_2$	Farringtonite	c	−4022.9	−3782.	237.6	1
			−3780.7	−3538.8	189.2	4
			−3790.3	−3548.4	189.2	6
			−3780.7	−3538.7	189.2	31
			−4022.9	−3782.3	237.6	119
$MgNH_4PO_4$		c		−1631.8		1
				−1624.9		4
$MgNH_4PO_4 \cdot 6H_2O$	Struvite	c	−3681.9	−3051.1		4
			−3681.9			31
$Mg(NO_3)_2$		c	−790.5	−589.1	164.0	6
			−790.6	−589.2	164.0	21
			−790.6	−589.4	164.0	31
$Mg_3(AsO_4)_2$		c	−3059.8	−2842.2	225.1	1
			−3092.8			31
$MgFe_2O_4$	Magnesioferrite	c	−1428.7	−1317.4	123.8	6

Formula	Description	State	ΔH_f° kJ	ΔG_f° kJ	S° J/deg	Source
			−1428.4	−1317.0	123.8	21
			−1428.4	−1317.1	123.8	31
			−1464.0	−1351.0	118.4	119
$MgAl_2O_4$	Spinel	c	−2315.0	−2190.2	80.6	6
			−2308.7	−2182.9		11
			−2299.3	−2174.9	80.6	21
			−2299.9	−2175.2	80.6	31
			−2296.0	−2171.2	80.6	119
$MgCr_2O_4$	Picrochromite	c			105.8	6
					105.8	11
			−1783.6	−1669.1	106.0	21
			−1783.6	−1668.9	106.0	31
			−1751.2		105.8	119
$MgSiO_3$	Clinoenstatite	c	−1549.1	−1467.9	67.9	6
			−1549.0	−1462.2	67.9	7
			−1549.3	−1462.3	67.9	11
			−1547.8	−1460.9	67.9	21
			−1549.0	−1462.1	67.7	31

Formula	Description	State	ΔH°_f kJ	ΔG°_f kJ	S° J/deg	Source
			−1547.8	−1460.9	67.9	47
			−1554.8	−1458.0	67.8	108
$MgSiO_3$	Enstatite	c	−1546.8	−1459.9	67.8	22
				−1460.0	67.7	44
					66.8	53
			−1544.7	−1457.4	66.3	108
Mg_2SiO_4	Forsterite	c	−2172.3	−2053.3	95.2	6
			−2177.5	−2058.6	95.2	7
			−2181.3	−2062.0	95.2	11
			−2170.4	−2051.3	95.2	21
			−2175.7	−2056.7	95.2	22
			−2174.0	−2055.1	95.1	31
				−2055.6		43
			−2170.4	−2051.3	95.2	47
					92.4	53
				−2052.5		83
			−2172.7	−2053.8	95.5	108
			−2177.9	−2059.2	95.0	119

Formula	Description	State	ΔH°_f kJ	ΔG°_f kJ	S° J/deg	Source
Mg_2SiO_4	Forsteritic spinel	c	−2147.5	−2022.6	75.3	11
$Mg_3Si_4O_{10}(OH)_2$	Talc	c	−5922.9	−5543.4	260.8	6
			−5916.0	−5536.6	260.8	7
			−5921.3	−5540.9	260.8	11
				−5523		13
			−5915.9	−5536.0	260.8	21
			−5903.3	−5523.7	260.8	22
			−5922.5	−5542.7	260.7	31
				−5524.6		43
			−5915.9	−5536.3	260.8	47
					293.6	53
			−6200.2	−5514.9	260.8	108
$Mg_3Si_2O_5(OH)_4$	Chrysotile	c		−4025		13
				−4037.2		14
			−4361.7	−4034.0	221.3	21
			−4364.4	−4037.0	221.3	22
			−4365.6	−4037.8	221.3	31
			−4361.7	−4034.0	221.3	47

Formula	Description	State	ΔH°_f kJ	ΔG°_f kJ	S° J/deg	Source
					220.2	53
				-4036.5	221.3	83
			-4482.0	-4032.4	221.3	108
$Mg_3Si_2O_5(OH)_4$	Serpentine	c	-4365.4	-4038.0	221.3	6
			-4365.3	-4040.8	230.0	7
			-4373.6	-4046.1	225.1	11
$Mg_{2.825}Si_2O_5(OH)_{3.65}$	Antigorite	c	-4201.4	-3890.6	212.0	22
				-3885.9		44
			-4199.8	-3889.0	212.1	108
$Mg_3Si_2O_5(OH)_4$	Antigorite	c			222.6	31
$Mg_7Si_8O_{22}(OH)_2$	Anthophyllite	c	-12090.9	-11364.6	534.7	6
			-12089.9	-11371.6	559.2	7
			-12125.7	-11396.1	523.3	11
			-12086.5	-11361.4	538.1	22
				-11344.2		44
			-12058.4	-11332.9	537.0	108
$Mg_4Si_6O_{15}(OH)_2 \cdot 6(H_2O)$	Sepiolite	c		-9251.7		15
			-10116.9	-9251.6	613.4	22

Formula	Description	State	ΔH_f° kJ	ΔG_f° kJ	S° J/deg	Source
$Mg_4Si_6O_{12}(OH)_8$	Sepiolite	c		-8539.5		8
$Mg_2Al_4SiO_{10}$	Sapphirine	c	-5278.0	-4981.6	228.5	11
$Mg_3Al_2Si_3O_{12}$	Pyrope	c	-6326.8	-5978.0	271.8	11
			-6284.6	-5920.8	222.0	21
			-6212.7			51
$Mg_2Al_4Si_5O_{18}$	Cordierite	c	-9175.2	-8665.0	407.1	11
			-9161.5	-8651.1	407.2	21
			-9134.5	-8624.4	407.2	22
				-8740.0		75
			-9224	-8715		84
			-9108.6	-8598.1	407.1	119
$Mg_2Al_4Si_5O_{18} \cdot H_2O$	Hydrous cordierite	c	-9437.7	-8875.7	466.2	22
$Mg_2Al_4Si_5O_{18} \cdot 0.6H_2O$	Hydrous cordierite	c		-8774		84
$Mg_4Al_4Si_2O_{10}(OH)_8$	Amesite	c	-9067.3	-8422.3	485.8	11
$Mg_5Al_2Si_3O_{10}(OH)_8$	7-Å clinochlore	c	-8841.6	-8188.5	445.6	22
$Mg_5Al_2Si_3O_{10}(OH)_8$	14-Å clinochlore	c	-8857.4	-8207.8	465.3	22
			-8883.	-8238.	476.1	84
$Mg_5Al_2Si_3O_{10}(OH)_8$	Clinochlore	c		-8260.5		75

Formula	Description	State	ΔH°_f kJ	ΔG°_f kJ	S° J/deg	Source
$Mg_5Al_2Si_3O_{10}(OH)_8$	Chlorite	c	−8901.6	−8257.3	483.0	11
$Mg_4Al_{10}Si_7O_{31}(OH)_4$	Yoderite	c	−17870.8	−16805.1	−820.8	11
$Mg_{0.167}Al_{2.33}Si_{3.67}O_{10}(OH)_2$	Montmorillonite	c	−5707.6	−5336.0	256.1	20
			−5710.0	−5332.5	236.0	120
$Mg_{0.2}(Si_{3.81}Al_{1.71}Fe^{3+}_{0.22}$ $Mg_{0.29})O_{10}(OH)_2$	Montmorillonite	c		−5254.3		41
$MgTiO_3$	Geikelite	c	−1571.1	−1482.2	74.6	6
			−1572.8	−1484.4	74.6	21
			−1572.8	−1484.5	74.5	119

- MANGANESE -

Formula	Description	State	ΔH°_f kJ	ΔG°_f kJ	S° J/deg	Source
Mn α	Metal	c	0	0	31.8	3
			0	0	32.0	6
			0	0	32.0	11
			0	0	32.0	21
Mn γ	Metal	c	1.5	1.4	32.3	3
			1.6	1.4	32.4	31

Formula	Description	State	ΔH_f° kJ	ΔG_f° kJ	S° J/deg	Source
Mn^{2+}		aq	−223.0	−227.6	−84.	1
			−220.5	−230.0	−66.9	6
			−220.7	−228.0	−73.6	21
			−220.8	−228.1	−73.6	31
				−228.0		100
			−241.4	−230.0	−66.9	119
Mn^{3+}		aq	−113.	−82.0		1
			−100.4	−85.4	−213.	6
			−100.4	−85.4	−213.4	119
MnO_4^-		aq	−542.7	−449.4	190.0	1
			−533.0	−440.3	196.2	6
			−541.4	−447.2	191.2	31
			−533.0	−440.3	196.2	119
MnO_4^{2-}		aq		−503.8		1
				−494.1		6
			−653.	−500.7	59.	31
				−494.1		119
$Mn(OH)^+$		aq	−446.0	−406.7	2.5	6

Formula	Description	State	ΔH°_f kJ	ΔG°_f kJ	S° J/deg	Source
			−450.6	−405.0	−17.	31
			−446.0	−406.7	2.5	119
$HMnO_2^-$		aq		−505.8		1
				−505.8		5
$MnCO_3$		aq	−895.0	−751.4	−136.8	3
$MnHCO_3^+$		aq		−820.		31
				−825.9		61
MnO	Manganosite	c	−384.9	−363.2	60.2	3
			−384.9	−362.8	59.7	6
			−385.1	−362.8	59.7	11
			−385.2	−362.9	59.7	21
			−385.2	−362.9	59.7	31
MnO_2 γ	Pyrolusite	c	−519.6	−464.8	53.1	1
			−520.1	−465.2	53.0	6
			−520.7	−465.8	53.0	11
			−520.0	−465.1	53.0	21
			−520.0	−465.1	53.0	31
				−465.7		100

Formula	Description	State	ΔH_f° kJ	ΔG_f° kJ	S° J/deg	Source
MnO_2	Birnessite ($MnO_{1.7}$ to MnO_2, contains hydroxyl)	c		-453.1		56
MnO_2	Nsutite ($MnO_{1.75}$ to MnO_2, contains hydroxyl)	c		-456.5		56
Mn_2O_3	Bixbyite	c	-971.1	-888.3	92.5	1
			-957.7	-879.9	110.4	6
			-959.0	-881.4	110.4	11
			-959.0	-881.1	110.5	21
			-959.0	-881.1	110.5	31
			-959.8	-889.9	110.4	119
Mn_2O_3 γ		c		-553.1		56
Mn_3O_4	Hausmannite	c	-1384.9	-1278.6	148.5	6
			-1386.6	-1280.0	148.5	11
			-1387.8	-1282.8	154.0	21
			-1387.8	-1283.2	155.6	31
			-1387.8	-1282.8		50
				-1281.1		56
$Mn(OH)_2$	Pyrochroite	c	-702.1	-616.7	81.6	6
				-615.6		56

Formula	Description	State	ΔH_f° kJ	ΔG_f° kJ	S° J/deg	Source
				−616.3		100
			−697.9	−614.6	88.3	119
$Mn(OH)_2$	Pptd.	am	−695.4	−615.0	99.2	31
$Mn(OH)_3$		c	−887.	−757.	99.6	1
MnOOH	Manganite	c	−615.3	−557.3		50
				−557.7		56
$MnCl_2$	Scacchite	c	−481.2	−440.5	118.2	6
			−481.3	−440.5	118.2	21
			−481.3	−440.5	118.2	31
			−472.8		117.2	119
MnS	Alabandite	c	−204.2	−208.8	78.2	3
			−213.8	−218.1	78.3	6
			−213.9	−218.2	78.2	21
			−213.4	−218.3	80.3	22
			−214.2	−218.4	78.2	31
MnS	Pptd.	c		−223.0		1
MnS	Pptd., pink	am	−213.8			31
MnS_2	Hauerite	c	−244.6	−232.2	54.0	6

Formula	Description	State	ΔH_f° kJ	ΔG_f° kJ	S° J/deg	Source
					99.9	21
$MnCO_3$	Rhodochrosite, natural	c	−894.1	−816.7	85.8	6
			−888.6	−815.5		11
			−889.3	−816.0	100.0	21
			−889.2	−816.1	100.0	22
			−894.1	−816.7	85.8	31
			−889.3	−815.0		50
				−818.8		65
$(Mn_{0.994}Fe_{0.005}Mg_{0.001})CO_3$	Rhodochrosite, natural	c	−891.9	−818.1	98.03	106
$MnCO_3$		c	−895.0	−817.6	85.6	3
$MnCO_3$	Pptd.	c	−887.	−813.0	99.6	1
			−883.2			31
$MnSO_4$		c	−1063.7	−956.0	112.1	3
			−1065.8	−958.0	112.1	6
			−1065.2	−957.3	112.1	21
			−1065.2	−957.4	112.1	31
$Mn_3(PO_4)_2$	Pptd.	c	−3226.	−2858.	299.6	1
			−3115.9			6

Formula	Description	State	ΔH°_f kJ	ΔG°_f kJ	S° J/deg	Source
			-3116.7			31
$MnWO_4$	Huebnerite	c	-1305.4	-1204.2	135.1	6
					132.5	21
			-1305.0			31
$MnSiO_3$	Rhodonite	c	-1265.7	-1185.3	89.1	3
			-1320.6	-1240.4	89.1	6
			-1320.3	-1240.2	102.5	11
			-1319.4	-1243.1	102.5	21
			-1320.9	-1240.5	89.1	31
			-1319.4	-1243.0	102.5	47
			-1320.7	-1240.8	89.1	119
Mn_2SiO_4	Tephroite	c	-1730.0	-1631.8	163.2	6
			-1729.7	-1632.4	163.2	11
			-1728.1	-1629.7	163.2	21
			-1730.5	-1632.1	163.2	31
			-1728.1	-1629.7	163.2	47
			-1730.3	-1630.0	154.4	119

Formula	Description	State	ΔH°_f kJ	ΔG°_f kJ	S° J/deg	Source
			- MERCURY -			
Hg	Metal	c	0	0	77.4	3
			0	0	76.0	6
			0	0	75.9	21
			0	0	76.0	22
			0	0	76.0	31
Hg		g	60.8	31.8	174.9	3
					174.9	6
			61.3	31.8	175.0	31
Hg^{2+}		aq	174.0	164.8	−22.6	1
			171.1	164.4	−32.2	6
			171.0	164.4	−32.0	21
			171.0	164.4	−32.2	31
			170.2		−36.0	85
			171.1	164.4	−32.2	119
Hg_2^{2+}		aq		152.1		1
			172.4	153.6	84.5	6

Formula	Description	State	ΔH°_f kJ	ΔG°_f kJ	S° J/deg	Source
			172.0	153.6	84.5	21
			172.4	153.5	84.5	31
			172.4	153.6	84.5	119
$Hg(OH)^+$		aq	−84.5	−52.3	71.1	6
			−84.5	−52.3	71.	31
			−84.5	−52.3	71.1	119
$Hg(OH)_2$		aq		−274.9		1
			−355.2	−274.9	142.	6
			−355.2	−274.8	142.	31
			−355.2	−274.9	142.2	119
$HHgO_2^-$		aq		−190.0		3
				190.3		31
$HgCl^+$		aq	−18.8	−5.4	75.3	6
			−18.8	−5.4	75.	31
			−18.8	−5.4	75.3	119
$HgCl_2$		aq	−216.3	−173.2	154.8	6
			−216.3	−173.2	155.	31
			−216.3	−173.2	154.8	119

Formula	Description	State	ΔH_f° kJ	ΔG_f° kJ	S° J/deg	Source
$HgCl_3^-$		aq	−388.7	−309.2	209.2	6
			−388.7	−309.1	209.	31
			−388.7	−309.2	209.2	119
$HgCl_4^{2-}$		aq		−450.6		1
			−554.0	−446.8	293	31
$HgBr_4^{2-}$		aq	−418.0	−368.2	351.	1
			−431.0	−371.1	310.	31
HgS_2^{2-}		aq		48.5		1
				41.8		6
				41.9		31
				41.8		119
HgO	Montroydite, red	c	−90.7	−58.5	72.0	3
			−90.7	−58.4	70.3	6
			−90.8	−58.5	70.3	21
			−90.8	−58.5	70.3	31
HgO	Yellow	c	−90.2	−58.4	73.2	3
			−90.5	−58.4	71.1	31
HgO	Hexagonal	c	−89.5	−58.2	73.6	31

Formula	Description	State	ΔH°_f kJ	ΔG°_f kJ	S° J/deg	Source
HgCl		g	79.	58.	260.2	3
			84.1	62.7	259.9	31
Hg_2Cl_2	Calomel	c	-264.9	-210.7	195.8	3
			-265.2	-210.8	192.5	6
			-265.2	-210.8	192.5	21
			-265.2	-210.7	192.5	31
$HgCl_2$		c	-230.1	-185.8	144.3	1
			-224.3	-178.6	146.0	6
			-224.3	-178.6	146.0	31
$HgBr_2$		c	-169.4	-147.4	155.6	1
			-170.7	-153.1	172.	31
Hg_2Br_2		c	-206.8	-178.7	213.0	3
				-181.1		6
			-206.9	-181.1	218.	31
HgI		g	138.	96.	280.7	3
			132.4	88.4	281.5	31
HgI_2	Coccinite, red	c	-105.4	-102.2	181.3	21
			-105.4	-101.7	180.	31

Formula	Description	State	ΔH°_f kJ	ΔG°_f kJ	S° J/deg	Source
Hg_2I_2	Yellow	c	−121.0	−111.3	239.3	3
				−111.0		6
			−121.3	−111.0	233.5	31
HgS	Cinnabar, red	c	−58.2	−48.8	77.8	3
			−58.2	−50.6	82.4	6
			−58.2	−50.6	82.5	21
			−53.3	−45.8	82.4	22
			−58.2	−50.6	82.4	31
HgS	Metacinnabar, black	c	−54.0	−46.2	83.3	3
			−53.6	−47.7	88.3	6
			−46.7	−43.3	96.2	21
			−43.7	−49.4	88.7	22
			−53.6	−47.7	88.3	31
HgH		g	242.9	220.1	219.3	3
			239.3	216.0	219.6	31
Hg_2CO_3		c		−442.7		1
			−553.5	−468.1	180.	31
$HgSO_4$		c	−704.2	−589.9	136.4	1

Formula	Description	State	ΔH_f° kJ	ΔG_f° kJ	S° J/deg	Source
			-707.5			31
Hg_2SO_4		c	-742.0	-623.9	200.7	3
			-743.1	-625.9	200.7	6
			-743.1	-625.8	200.7	31
Hg_2CrO_4		c		-651.6		1
				-617.6		6

- MOLYBDENUM -

Formula	Description	State	ΔH_f° kJ	ΔG_f° kJ	S° J/deg	Source
Mo	Metal	c	0	0	28.6	3
			0	0	28.5	6
			0	0	28.7	21
			0	0	28.7	31
					28.6	119
Mo^{3+}		aq		-57.7		12
				-57.7		54
MoO_4		aq	-725.9	-644.	167.4	1
	unspc. aq. soln.		-661.1			31

Molybdenum

Formula	Description	State	ΔH_f° kJ	ΔG_f° kJ	S° J/deg	Source
MoO_4^{2-}		aq	-997.9	-838.0	33.0	6
				-859.5		12
			-997.9	-836.3	27.2	31
			-997.9	-836.4	27.2	80
			-997.9	-838.0	33.0	119
$HMoO_4^-$		aq		-866.6		6
				-893.7		12
				-870.6		54
				-866.6		119
H_2MoO_4		aq		-950.		1
				-877.1		6
	unspc. aq. soln.		-1007.5			31
				-877.1		119
MoO_2		c	-589.1	-534.3	46.3	6
			-589.1	-532.0	46.3	11
				-502.1		12
			-587.8	-533.0	50.0	21
			-588.9	-533.0	46.3	31

Formula	Description	State	ΔH°_f kJ	ΔG°_f kJ	S° J/deg	Source
			-587.4		46.3	119
MoO_3	Molybdite	c	-754.5	-677.6	78.2	3
			-745.2	-667.8	77.7	6
			-745.2	-668.0	77.7	21
			-745.1	-668.0	77.7	31
			-745.2		77.7	119
$MoO_3 \cdot H_2O$		c		-1187.0		12
H_2MoO_4		c	-1046.0			31
				-1187.0		54
MoS_2	Molybdenite	c	-232.2	-225.1	63.2	3
			-234.9	-226.1	62.6	6
			-306.3	-297.4	62.6	21
			-235.1	-225.9	62.6	31
MoS_3		c	-256.1	-241.0	75.	1

Formula	Description	State	ΔH°_f kJ	ΔG°_f kJ	S° J/deg	Source
- NICKEL -						
Ni	Metal	c	0	0	30.1	3
			0	0	29.9	6
			0	0	29.9	21
			0	0	29.9	22
			0	0	29.9	31
Ni^{2+}		aq	−64.0	−48.2		1
			−53.6	−45.1	−129.0	6
			−54.0	−45.6	−129.0	21
			−54.0	−45.6	−128.9	31
			−54.0	−45.6	−128.9	80
			−53.6	−45.1	−129.0	119
$Ni(OH)^+$		aq		−220.0		6
			−287.9	−227.6	−71.	31
				−230.1		32
				−220.0		119
$HNiO_2^-$		aq		−349.2		12

Formula	Description	State	ΔH°_f kJ	ΔG°_f kJ	S° J/deg	Source
NiO	Bunsenite	c	−239.7	−211.6	38.0	6
			−239.7	−211.6	38.1	11
			−239.7	−211.6	38.0	21
			−239.7	−211.6	38.0	22
			−239.7	−211.7	38.0	31
			−239.7	−211.6	38.0	119
NiO_2		c		−198.7		1
Ni_3O_4		c		−711.9		62
$NiO_2 \cdot 2H_2O$		c		−689.5		12
$Ni_2O_3 \cdot H_2O$		c		−706.9		12
$Ni_3O_4 \cdot 2H_2O$		c		−1186.3		12
$Ni(OH)_2$		c	−538.1	−453.1	79.	3
				−459.1		6
			−529.7	−447.2	88.	31
$Ni(OH)_3$		c	−678.2	−541.8	81.6	1
	Pptd.		−669.			31
$NiCl_2$		c	−305.4	−259.2	97.6	6
			−305.3	−259.0	97.7	21

Formula	Description	State	ΔH°_f kJ	ΔG°_f kJ	S° J/deg	Source
			−305.3	−259.0	97.6	31
NiS α	Millerite	c		−74.0		1
			−82.0	−79.5	53.0	6
			−84.9	−86.2	66.1	21
			−82.0	−79.5	53.0	31
NiS γ		c		−114.2		1
	Pptd.		−77.4			31
Ni_3S_2	Heazlewoodite	c	−202.9	−197.1	133.9	6
			−202.9	−197.1	133.9	21
			−202.9	−197.1	133.9	31
$NiCO_3$		c	−664.0	−615.0	91.6	1
			−689.1	−612.1	85.5	6
				−612.5		31
$NiSO_4$		c	−891.2	−773.6	77.8	3
			−872.9	−763.2	103.8	6
			−872.9	−759.7	92.	31
$NiSO_4 \cdot 6H_2O$ α	Retgersite, green, tetragonal	c	−2698.6			3
			−2682.6			6

Formula	Description	State	ΔH_f° kJ	ΔG_f° kJ	S° J/deg	Source
			−2682.8	−2224.5	334.5	21
			−2682.8	−2224.6	334.5	31
$NiSO_4 \cdot 6H_2O$ β	Blue, monoclinic	c	−2688.2	−2221.7	305.8	3
			−2672.3			31
$NiSO_4 \cdot 7H_2O$	Morenosite	c	−2976.1	−2462.0	378.9	6
			−2976.3	−2461.7	378.9	21
			−2976.3	−2461.8	378.9	31
$NiFe_2O_4$	Trevorite	c	−1077.0	−968.9	131.8	6
			−1081.2	−972.9	131.8	21
			−1081.1	−973.1	131.8	31
$NiAl_2O_4$		c	−1939.7	−1819.2	92.5	6
			−1915.9			31
$NiSiO_3$		c		−1128.0		24
Ni_2SiO_4	Olivine	c	−1396.5	−1289.0	128.1	110
Ni_2SiO_4	Spinel	c		−1281	124.1	110
Ni_2SiO_4		c	−1429.7	−1317.1	111.3	6
			−1398.9	−1286.5	111.7	119

Formula	Description	State	ΔH°_f kJ	ΔG°_f kJ	S° J/deg	Source
			- NIOBIUM -			
Nb	Metal	c	0	0	34.7	3
			0	0	36.4	6
			0	0	36.4	21
			0	0	36.4	31
Nb^{3+}		aq		-318.		1
NbO		c		-378.6		12
			-419.7	-391.9	46.0	21
			-405.8	-378.6	48.1	31
NbO_2		c			54.5	6
				-736.		12
			-795.0	-739.2	54.5	21
			-796.2	-740.5	54.5	31
Nb_2O_4		c	-1622.6	-1516.3	122.2	1
Nb_2O_5		c	-1897.4	-1764.0	137.2	6
				-1765.6		12
			-1899.5	-1765.8	137.3	21

Formula	Description	State	ΔH°_f kJ	ΔG°_f kJ	S° J/deg	Source
	High temp. form		−1899.5	−1766.0	137.2	31

- NITROGEN -

Formula	Description	State	ΔH°_f kJ	ΔG°_f kJ	S° J/deg	Source
N_2		g	0	0	191.5	3
			0	0	191.5	6
			0	0	191.6	21
			0	0	191.6	31
N_2		aq	−10.5	18.2	95.4	6
				12.5		12
			−10.5	18.2	95.4	119
NO_2^-		aq	−106.3	−34.5	125.1	1
			−104.6	−3.7	140.2	6
			−104.6	−32.2	123.0	31
			−104.6	−3.7	140.2	119
HNO_2		aq	−119.2	−50.6	135.6	31
$N_2O_2^{2-}$		aq	−10.8	138.1	27.6	3
	unspc. aq. soln.		−17.2			31

Formula	Description	State	ΔH_f° kJ	ΔG_f° kJ	S° J/deg	Source
NO_3^-		aq	−206.6	−110.6	146.4	1
			−207.4	−111.4	146.8	6
			−207.4	−111.5	146.9	21
			−205.0	−108.7	146.4	31
			−207.4	−111.4	146.8	119
HNO_3		aq	−206.6	−110.5	146.4	3
	see preface		−207.4	−111.2	146.4	31
NH_3		aq	−80.8	−26.6	110.0	3
			−80.3	−26.6	111.0	21
			−80.3	−26.5	111.3	31
NH_4^+		aq	−132.8	−79.5	112.8	3
			−132.5	−78.7	111.3	6
			−133.3	−79.4	111.2	21
			−132.5	−79.3	113.4	31
			−132.5	−78.7	111.3	119
NH_2OH		aq	−90.8	−23.4	167.	1
	unspc. aq. soln.		−98.3			31
$NH_2(OH)_2^+$		aq		−56.6		12

Formula	Description	State	ΔH_f° kJ	ΔG_f° kJ	S° J/deg	Source
NH_4OH		aq	−366.7	−263.8	179.9	1
			−366.1	−263.2	179.5	6
			−366.1	−263.6	181.2	31
			−366.1	−263.2	179.5	119
NO		g	90.4	86.7	210.6	3
			90.2	86.6	210.6	6
			90.2	86.6	210.8	31
			90.3	86.7	210.6	119
NO_2		g	33.8	51.8	240.4	3
			33.2	51.3	240.0	6
			33.1	51.2	240.1	21
			33.2	51.3	240.1	31
			33.8	51.8		119
N_2O		g	81.5	103.6	220.0	3
			82.0	104.2	219.8	31
N_2O_4		g	9.7	98.3	304.3	3
			9.2	97.9	304.3	31
HNO_3		1	−173.2	−79.9	155.6	3

Formula	Description	State	ΔH°_f kJ	ΔG°_f kJ	S° J/deg	Source
			−174.1	−80.8	155.6	6
			−174.1	−80.7	155.6	31
NOCl		g	52.6	66.4	263.6	3
			51.7	66.1	261.7	31
NOBr		g	81.8	82.4	272.6	3
			82.2	82.4	273.7	31
NH_3		g	−46.2	−16.6	192.5	3
			−46.1	−16.5	192.3	6
			−45.9	−16.4	192.8	21
			−46.1	−16.4	192.4	31
NH_4Cl	Sal ammoniac	c	−314.4	−203.0	94.6	6
			−315.2	−203.8	95.0	21
			−314.4	−202.9	94.6	31
$(NH_4)_2SO_4$	Mascagnite	c	−1180.0	−901.1	220.1	6
			−1180.8	−901.7	220.1	21
			−1180.8	−901.7	220.1	31
NH_4HSO_4		c	−1027.0			6
			−1027.0			31

Formula	Description	State	ΔH°_f kJ	ΔG°_f kJ	S° J/deg	Source
NH_4NO_3		c	-365.6	-183.9	150.9	6
			-365.6	-183.8	151.1	21
			-365.6	-183.9	151.1	31
NH_4VO_3		c	-1053.1			6
			-1053.1	-888.1	140.6	31
				-928.0		35

- OXYGEN -

Formula	Description	State	ΔH°_f kJ	ΔG°_f kJ	S° J/deg	Source
O_2		g	0	0	205.0	3
			0	0	205.0	6
			0	0	205.0	11
			0	0	205.2	21
			0	0	205.0	22
			0	0	205.1	31
O_2		aq	-11.7	16.3	110.9	6
			-11.7	16.4	110.9	31
				16.5		55

Formula	Description	State	ΔH°_f kJ	ΔG°_f kJ	S° J/deg	Source
			-11.7	16.3	110.9	119
O_2^-		aq		54.		1
OH^-		aq	-229.9	-157.3	-10.5	1
			-230.0	-157.3	-10.8	6
			-230.0	-157.3	-10.7	21
			-230.0	-157.2	-10.8	31
			-230.0	-157.3	-10.8	119
HO_2^-		aq		-65.3		1
			-160.3	-67.3	23.8	31
H_2O_2		aq	-191.1	-131.7		1
			-191.2	-134.0	143.9	31
H_2O	Steam	g	-241.8	-228.6	188.7	3
			-241.8	-228.6	188.7	6
			-241.8	-228.6	188.7	11
			-241.8	-228.6	188.7	21
			-241.8	-228.6	188.7	22
			-241.8	-228.6	188.8	31
H_2O		l	-285.8	-237.2	69.9	3

Formula	Description	State	ΔH°_f kJ	ΔG°_f kJ	S° J/deg	Source
			-285.8	-237.2	69.9	6
			-285.8	-237.2	69.9	11
			-285.8	-237.1	70.0	21
			-285.8	-237.2	69.9	22
			-285.8	-237.1	69.9	31
H_2O_2		l	-187.8	-120.4	109.6	31
				-120.5		55

– PALLADIUM –

Formula	Description	State	ΔH°_f kJ	ΔG°_f kJ	S° J/deg	Source
Pd	Metal	c	0	0	37.2	3
			0	0	37.9	6
			0	0	37.8	21
			0	0	37.6	31
Pd^{2+}		aq		190.4		1
				176.6		6
			149.0	176.5	-184.	31
				176.6		119

Formula	Description	State	ΔH_f° kJ	ΔG_f° kJ	S° J/deg	Source
PdO		c	-85.4	-60.2	55.2	1
			-118.8			6
			-85.4			31
PdO_3		c		100.8		12
$Pd(OH)_2$		c	-385.3	-301.	90.8	1
	Pptd.		-395.0			31
$Pd(OH)_4$		c	-708.8	-528.0	103.3	1
	Pptd.		-715.9			31
Pd_2H		c		-4.6		12
			-19.7			31

- PHOSPHORUS -

Formula	Description	State	ΔH_f° kJ	ΔG_f° kJ	S° J/deg	Source
P	White	c	0	0	44.4	3
			0	0	41.1	6
			0	0	22.8	21
			0	0	41.1	31
P	Red	c	-18.4	-13.8	29.3	1

Formula	Description	State	ΔH_f° kJ	ΔG_f° kJ	S° J/deg	Source
			−17.4	−11.9	22.8	6
			−17.6	−12.1	22.8	31
P	Black	c	−43.1			3
			−38.9	−33.5	22.7	6
			−39.3			31
P_2		g	141.5	102.9	218.1	3
			144.3	103.7	218.1	31
P_4		g	54.9	24.4	279.9	3
				24.4		12
			58.9	24.4	280.0	31
PO_4^{3-}		aq	−1284.1	−1025.5	−218.	3
			−1277.4	−1018.8	−220.3	6
			−1277.0	−1019.0	−222.0	21
			−1277.4	−1018.7	−222.	31
			−1277.4	−1018.8	−221.8	119
$P_2O_6^{4-}$		aq		−1513.4		12
HPO_3^{2-}		aq	−978.2	−811.7		1
			−969.0			6

Formula	Description	State	ΔH°_f kJ	ΔG°_f kJ	S° J/deg	Source
	unspc. aq. soln.		−969.0			31
			−969.0			119
HPO_4^{2-}		aq	−1298.7	−1094.1	−36.0	3
			−1292.1	−1089.3	−33.5	6
			−1292.1	−1089.1	−33.5	31
			−1292.1	−1089.3	−33.5	119
$HP_2O_6^{3-}$		aq		−1570.2		12
$H_2PO_2^{-}$		aq		−512.1		1
	unspc. aq. soln.		−613.8			31
$H_2PO_3^{-}$		aq		−846.6	79.	1
			−969.4			6
	unspc. aq. soln.		−969.4			31
			−969.4			119
$H_2PO_4^{-}$		aq	−1302.5	−1135.1	89.1	3
			−1296.3	−1130.4	90.4	6
			−1296.3	−1130.3	90.4	31
			−1296.3	−1130.4	90.4	119
$H_2P_2O_6^{2-}$		aq		−1611.7		12

Formula	Description	State	ΔH_f° kJ	ΔG_f° kJ	S° J/deg	Source
H_3PO_2		aq	−609.2	−523.4	159.	1
H_3PO_3		aq	−971.5	−856.9	167.	1
			−964.8			6
	unspc. aq. soln.		−964.8			31
			−964.8			119
H_3PO_4		aq	−1289.5	−1147.2	176.1	1
			−1288.3	−1142.6	158.2	6
			−1288.3	−1142.5	158.2	31
			−1288.3	−1142.6	158.2	119
$H_3P_2O_6^-$		aq		−1627.6		12
$H_4P_2O_5$		aq		−1640.		1
PO		g	−12.1	−41.2	222.8	21
			−28.5	−51.9	222.8	31
P_2O_5	Orthorhombic	c	−1520.7			6
			−1504.9	−1372.8	115.5	21
P_2O_5	Hexagonal	c	−1492.0	−1348.9	114.3	6
$(P_2O_5)_2$	Dimeric	c	−3009.8	−2745.6	231.0	21
P_4O_{10}	Hexagonal	c	−3010.0	−2723.6	228.8	4

Formula	Description	State	ΔH_f° kJ	ΔG_f° kJ	S° J/deg	Source
			-2984.0	-2697.7	228.9	31
PCl_3		g	-306.4	-286.3	311.7	3
			-287.0	-267.8	311.8	31
PCl_5		g	-398.9	-324.6	352.7	3
			-374.9	-305.0	364.6	31
PH_3		g	9.2	18.2	210.0	3
			5.4	13.4	210.2	31
H_3PO_4		c	-1279.0	-1119.2	110.5	6
			-1266.9	-1112.3	110.5	21
			-1279.0	-1119.1	110.5	31
H_3PO_4		l	-1254.2	-1111.7	150.8	21
			-1266.9			31

- PLATINUM -

Formula	Description	State	ΔH_f° kJ	ΔG_f° kJ	S° J/deg	Source
Pt	Metal	c	0	0	41.8	3
			0	0	41.6	6
			0	0	41.6	21

Formula	Description	State	ΔH°_f kJ	ΔG°_f kJ	S° J/deg	Source
Pt^{2+}			0	0	41.6	31
		aq		229.3		1
				185.8		6
				254.8		31
				185.8		119
$Pt(OH)_2$		c	−364.8	−285.3	110.9	3
			−351.9			31
PtS	Cooperite	c	−87.0	−90.4	84.5	1
			−81.6	−76.1	55.1	6
			−82.4	−76.9	55.1	21
			−81.6	−76.1	55.1	31
PtS_2		c	−116.3	−107.1	74.5	1
			−108.8	−99.6	74.7	6
			−108.8	−99.6	74.7	31

Formula	Description	State	ΔH_f° kJ	ΔG_f° kJ	S° J/deg	Source
			- POTASSIUM -			
K	Metal	c	0	0	63.6	3
			0	0	64.6	6
			0	0	64.2	11
			0	0	64.7	21
			0	0	64.2	31
K^+		aq	−251.2	−282.3	106.7	3
			−252.3	−282.7	101.2	6
			−252.2	−282.5	101.0	21
			−252.4	−283.3	102.5	31
			−252.3	−282.7	101.2	119
KSO_4^-		aq		−1029.7		2
			−1157.3	−1030.8	147.3	6
			−1157.4	−1032.7	151.0	31
			−1157.3	−1030.8	147.3	119
K_2O		c	−361.5	−318.8	87.0	1
			−361.5			6

Formula	Description	State	ΔH°_f kJ	ΔG°_f kJ	S° J/deg	Source
			-361.5	-318.8	87.0	11
			-363.2	-322.1	94.1	21
			-363.2	-322.4	94.1	22
			-361.5			31
KO_2		c	-284.5	-240.6	122.5	21
			-284.9	-239.4	116.7	31
KOH		c	-425.8	-374.5	59.4	1
			-424.8			6
				-380.2		9
			-424.7	-378.9	78.9	21
			-424.8	-379.1	78.9	31
KCl	Sylvite	c	-435.9	-408.3	82.7	3
			-436.7	-408.9	82.4	6
			-436.5	-408.6	82.6	21
			-436.7	-408.9	82.6	22
			-436.7	-409.1	82.6	31
KBr		c	-393.8	-380.7	95.9	3
			-393.7	-380.4	95.9	6

Formula	Description	State	ΔH°_f kJ	ΔG°_f kJ	S° J/deg	Source
			-393.5	-380.1	95.9	21
			-393.8	-380.7	95.9	31
KI		c	-327.6	-322.3	104.3	3
			-330.3	-324.9	104.2	6
			-327.9	-324.9	106.3	31
K_2S		c	-418.	-404.2	111.3	1
			-380.7	-364.0	105.	31
K_2CO_3		c	-1146.1	-1069.0	140.6	1
			-1149.8	-1064.0	155.5	6
				-1060.6		24
			-1151.0	-1063.5	155.5	31
			-1146.1	-1060.6	149.4	119
K_2SO_4	Arcanite	c	-1437.7	-1320.0	175.6	6
			-1437.7	-1319.7	175.6	21
			-1437.8	-1321.4	175.6	31
$KHSO_4$	Mercallite	c	-1160.9			6
			-1160.6	-1031.3	138.1	31
				-1035.0		72

Formula	Description	State	ΔH°_f kJ	ΔG°_f kJ	S° J/deg	Source
$KAl(SO_4)_2$		c	−2470.0	−2240.0	204.6	6
			−2470.2	−2240.8	204.6	21
			−2470.2	−2240.0	204.6	31
$KAl_3(SO_4)_2(OH)_6$	Alunite	c			328.0	6
					328.0	21
			−5169.8	−4659.3	328.0	22
			−5169.8		328.0	30
				−4659.3		48
			−5169.8	−4659.3	318.4	68
$K_3Al_5(PO_4)_8H_6 \cdot 18H_2O$	Potassium taranakite	c	−18908.2	−16760.8	1328.1	4
			−18919.2	−17417.3	3496.2	31
KNO_3	Niter	c	−492.7	−393.1	132.9	3
			−494.5	−394.8	133.1	6
			−494.5	−394.5	133.1	21
			−494.6	−394.9	133.0	31
K_2SiO_3		c	−1541.7			6
			−1558.1	−1464.6	138.1	11
					146.0	31

Formula	Description	State	ΔH_f° kJ	ΔG_f° kJ	S° J/deg	Source
			−1558.5	−1465.0	138.1	119
$K_2Si_2O_5$		c	−2482.8	−2335.2	182.0	11
					190.6	31
$K_2Si_2O_5$ γ		c	−2483.2	−2335.6	182.0	119
$K_2Si_4O_9$		c	−4330.0	−4074.0	265.7	11
$K_2Si_4O_9$ β		c	−4330	−4095.4	265.7	119
$KAlSiO_4$	Kalsilite	c	−2108.7	−1993.0	133.0	6
			−2087.6	−1971.9	133.3	11
			−2131.4	−2015.6	133.3	22
$KAlSiO_4$	Kaliophilite	c	−2121.9	−2006.0	133.3	21
			−2121.3	−2005.3	133.1	31
			−2121.9	−2006.0	133.3	47
$KAlSi_2O_6$	Leucite	c	−3019.8	−2852.5	184.1	6
			−3016.6			11
			−3038.6	−2875.9	200.2	21
			−3034.2	−2871.4	200.0	31
			−3021.3	−2853.9	184.2	36
			−3038.7	−2875.9	200.2	47

Formula	Description	State	ΔH_f° kJ	ΔG_f° kJ	S° J/deg	Source
					183.8	53
$KAlSi_3O_8$	Sanidine	c	−3951.3	−3733.2	238.2	6
			−3948.5	−3730.5	238.2	11
			−3959.7	−3739.9	232.9	31
			−3959.5	−3739.7	232.9	47
					238.7	53
$KAlSi_3O_8$	High sanidine	c	−3959.6	−3739.8	232.9	21
			−3960.3	−3739.4	228.2	22
$KAlSi_3O_8$	Microcline	c	−3959.2	−3735.6	219.7	6
			−3956.4	−3732.8	219.5	11
			−3967.7	−3742.3	214.2	21
			−3968.1	−3742.9	214.2	31
			−3961.1	−3737.4	219.8	36
			−3967.7	−3742.3	214.2	47
					219.4	53
$KAlSi_3O_8$	Maximum microcline	c	−3971.4	−3746.2	213.9	22
$KAlSi_3O_8$	Adularia	c	−3951.0			11
$KAlSi_3O_8$		am	−3904.8	−3694.3	263.4	6

Formula	Description	State	ΔH°_f kJ	ΔG°_f kJ	S° J/deg	Source
			-3902.0	-3692.0	264.8	11
			-3914.7	-3703.5	261.6	21
			-3917.5			31
			-3914.7	-3703.5	261.6	47
$KAl_3Si_3O_{10}(OH)_2$	Muscovite	c		-5439.		2
			-5947.1	-5566.4	288.7	6
				-5609.1		8
			-5983.2	-5600.4	288.7	11
			-5976.7	-5600.7	306.4	21
			-5972.3	-5591.1	287.8	22
			-5984.4	-5608.4	306.3	31
			-5950.0	-5569.0	288.8	36
			-5976.7	-5600.7	306.4	47
				-5601.5		99
$K_{0.33}Al_{2.33}Si_{3.67}O_{10}(OH)_2$	K-montmorillonite	c	-5727.6	-5353.8	265.3	11
			-5730.3	-5356.4	265.4	20
$K_{0.38}Al_{2.38}Si_{3.62}O_{10}(OH)_2 \cdot H_2O$	K-montmorillonite	c	-6088.5	5654.2	297.1	11
$KFe_3(AlSi_3O_{10})(OH)_2$	Annite	c		-4808.7		8

Formula	Description	State	ΔH°_f kJ	ΔG°_f kJ	S° J/deg	Source
			−5208.0	−4849.6	389.2	11
				−4794.0		17
			−5155.5	−4799.7	398.3	22
			−5443.8			31
			−5320.0	−4850.3		36
$KFe^{3+}_{0.3}Fe^{2+}_{2.7}(AlSi_3O_{12})H_{1.7}$	Annite	c		−4818.7		8
				−4804.1		18
$KMg_{0.5}Al_2Si_{3.5}O_{10}(OH)_2$	Phengite	c	−5910.4	−5529.2	285.3	11
$KMg_3(AlSi_3O_{10})(OH)_2$	Phlogopite	c		−5903.2		8
			−6256.6	−5872.6	319.6	11
				−5888.6		17
					319.7	21
			−6226.1	−5841.6	318.4	22
				−5831.8		31
					315.6	53
				−5813.7		75
			−6562.4	−5878.1	319.8	77
$KMg_3(AlSi_3O_{10})F_2$	Fluor-phlogopite	c	−6371.4	−6026.2	317.6	6

Formula	Description	State	ΔH_f° kJ	ΔG_f° kJ	S° J/deg	Source
			-6392.9	-6053.1	336.3	21
			-6383.9	-6044.0	336.4	31
			-6392.9	-6053.1	336.3	47
					311.4	53
$K_{0.6}Mg_{0.25}Al_{2.3}Si_{3.5}O_{10}(OH)_2$	Illite	c	-5819.2	-5443.3	277.8	11
			-5821.9	-5445.9	278.0	20
$(K_{0.56}Na_{0.04})(Mg_{0.24}Al_{1.9})$ $(Si_{3.22}Al_{0.78})O_{10}(OH)_2$	Grundy illite	c		-5534.2		49
$K_{0.64}(Al_{1.54}Fe_{0.29}Mg_{0.19})$ $(Si_{3.51}Al_{0.49})O_{10}(OH)_2$	Fithian illite	c		-5316.2		40
$(K_{0.59}Na_{0.02}Ca_{0.01})(Al_{1.54}Fe_{0.29}^{3+}$ $Mg_{0.23})(Si_{3.47}Al_{0.53})O_{10}(OH)_2$	Fithian illite	c		-5364.3		46
$(K_{0.6}Na_{0.05}Ca_{0.07})(Mg_{0.04}Al_{1.65})$ $(Si_{3.46}Al_{0.54})O_{10}(OH)_2$	Fithian illite	c		-5521.6		49
$K_{0.53}(Al_{1.66}Fe_{0.20}^{3+}Mg_{0.13})$ $(Si_{3.62}Al_{0.39})O_{10}(OH)_2$	Beaver's Bend illite	c		-5303.6		40
$(K_{0.60}Na_{0.04})(Al_{1.43}Fe_{0.42}^{3+}Mg_{0.16})$ $(Si_{3.48}Al_{0.52})O_{10}(OH)_2$	Beaver's Bend illite	c		-5232.1		46

Formula	Description	State	ΔH°_f kJ	ΔG°_f kJ	S° J/deg	Source
$K_{0.59}(Al_{1.58}Fe_{0.24}Mg_{0.15})$ $(Si_{3.65}Al_{0.35})O_{10}(OH)_2$	Goose Lake illite	c		−5294.0		40
$(K_{0.59}Na_{0.03}Ca_{0.03})(Mg_{0.34}Al_{1.69})$ $(Si_{3.57}Al_{0.43})O_{10}(OH)_2$	Rock Island illite	c		−5469.7		49
$(K_{0.69}Na_{0.03}Ca_{0.05})(Mg_{0.4}Al_{1.6})$ $(Si_{3.58}Al_{0.42})O_{10}(OH)_2$	Marblehead illite	c		−5484.4		49

− RADIUM −

Formula	Description	State	ΔH°_f kJ	ΔG°_f kJ	S° J/deg	Source
Ra		c	0	0	71.	31
Ra^{2+}		aq	−527.6	−561.5	54.	31
$RaOH^+$		aq	−753.0	−720.9	67	94
$RaCl^+$		aq	−692.6	−692.2	117	94
$RaCO_3^o$		aq	−1200.3	−1103.6	67	94
$RaSO_4^o$		aq	−1431.7	−1321.7	144.3	94
RaO		c	−523.			31
$RaCl_2 \cdot 2H_2O$		c	−1464.	−1302.8	213.	31
$RaCO_3$		c	−1216.4	−1136.8	117	94

Formula	Description	State	ΔH°_f kJ	ΔG°_f kJ	S° J/deg	Source
$RaSO_4$		c	−1471.1	−1365.6	138	31
			−1476.4	−1364.6	138	94

- RUBIDIUM -

Formula	Description	State	ΔH°_f kJ	ΔG°_f kJ	S° J/deg	Source
Rb	Metal	c	0	0	69.4	3
			0	0	76.7	6
			0	0	76.8	21
			0	0	76.8	31
Rb^+		aq	−246.4	−282.2	124.3	3
			−251.1	−283.8	121.3	6
			−251.1	−291.7	120.5	21
			−251.2	−284.0	121.5	31
RbO_2		c	−278.7			31
Rb_2O		c	−330.1	−290.8	109.6	1
			−330.1			6
			−339.			31
Rb_2O_2		c	−425.5	−349.8	103.8	1

Formula	Description	State	ΔH°_f kJ	ΔG°_f kJ	S° J/deg	Source
			−472.			31
Rb_2O_3		c	−488.3	−386.6	105.8	1
Rb_2O_4		c	−528.0	−395.8	116.3	1
RbOH		c	−413.8	−364.4	70.7	1
			−414.2			6
			−418.2			31
Rb_2S		c	−348.1	−337.2	133.9	1
			−360.7			31
RbH		c		−30.5		12
			−52.3			31
Rb_2CO_3		c	−1128.0	−1043.1	97.5	1
			−1136.0	−1051.0	181.3	31
			−1128.0	−1044.3	171.1	119

Formula	Description	State	ΔH°_f kJ	ΔG°_f kJ	S° J/deg	Source
			- SCANDIUM -			
Sc	Metal	c	0	0	33.	1
			0	0	34.6	6
			0	0	34.6	21
			0	0	34.6	31
Sc^{3+}		aq	−622.6	−601.2	−234.	1
			−631.9	−601.2	−264.4	6
			−614.2	−586.6	−255.	31
$Sc(OH)^{2+}$		aq		−812.1		6
				−810.4		12
			−861.5	−801.2	−134.	31
Sc_2O_3		c	−1908.6	−1819.4	77.0	6
			−1908.8	−1819.4	77.0	21
			−1908.8	−1819.4	77.0	31
$Sc(OH)_3$		c		−1228.0		1
			−1375.3	−1242.6	93.3	6
			−1363.6	−1233.3	100.	31

Formula	Description	State	ΔH_f° kJ	ΔG_f° kJ	S° J/deg	Source
				– SELENIUM –		
Se	Gray, hexagonal	c	0	0	41.8	3
			0	0	42.1	6
			0	0	42.3	21
			0	0	42.4	31
Se		g	202.4	162.2	176.6	3
			227.1	187.0	176.7	31
Se_2		g	138.6	88.5	252.0	3
					252.0	6
			146.0	96.2	252.0	31
Se^{2-}		aq	132.2	155.6	83.7	3
			64.0	129.3	−46.0	6
				129.0		21
				129.3		31
			64.0	129.3	−46.0	119
SeO_3^{2-}		aq	−512.1	−373.8	16.3	3
			−509.2	−363.8	−7.1	6

Formula	Description	State	ΔH°_f kJ	ΔG°_f kJ	S° J/deg	Source
			−509.2	−369.8	13.	31
			−509.2	−363.8	−7.1	119
SeO_4^{2-}		aq	−607.9	−441.1	23.8	3
			−599.1	−441.4	54.0	6
			−599.1	−441.3	54.0	31
			−599.1	−441.4	54.0	119
HSe^-		aq	102.9	98.6	177.0	3
			15.9	43.9	79.5	6
			15.9	44.0	79.	31
			15.9	43.9	79.5	119
H_2Se		aq	75.7	77.0	166.9	3
			19.2	22.2	163.6	6
			19.2	22.2	163.6	31
			19.2	22.2	163.6	119
$HSeO_3^-$		aq	−516.7	−411.3	127.2	3
			−514.5	−411.3	134.3	6
			−514.6	−411.5	135.1	31
			−514.5	−411.3	134.3	119

Formula	Description	State	ΔH°_f kJ	ΔG°_f kJ	S° J/deg	Source
$HSeO_4^-$		aq	−598.7	−452.7	92.0	3
			−576.1	−450.9	162.8	6
			−581.6	−452.2	149.4	31
			−576.1	−450.5	162.8	119
H_2SeO_3		aq	−512.1	−425.9	191.2	3
			−507.5	−426.2	207.9	6
			−507.5	−426.1	207.9	31
			−507.5	−426.2	207.9	119
H_2SeO_4		aq	−607.9	−441.1	23.8	3
SeO_2		c	−230.1	−173.6	56.9	1
			−225.4	−171.5	66.7	6
			−225.4			31
H_2Se		g	85.8	71.1	221.3	3
			29.7	15.9	218.9	6
			29.7	15.9	219.0	31
SeF_6		g	−1029.	−929.	314.2	3
			−1116.9	−1016.5	313.8	6
			−1117.	−1017.	313.9	31

Formula	Description	State	ΔH°_f kJ	ΔG°_f kJ	S° J/deg	Source

— SILICON —

Formula	Description	State	ΔH°_f kJ	ΔG°_f kJ	S° J/deg	Source
Si	Metal	c	0	0	18.7	3
			0	0	18.8	6
			0	0	18.8	7
			0	0	18.9	11
			0	0	18.8	21
			0	0	18.8	31
Si		g	368.4	323.9	167.9	3
			455.6	411.3	168.0	31
H_4SiO_4		aq	−1462.1	−1309.9	179.5	6
			−1460.7	−1308.4	179.5	7
				−1307.5		13
				−1307.9		14
			−1456.3		191.8	20
			−1460.0	−1308.0	180.0	21
				−1307.5		28
			−1468.6	−1316.6	180.	31

Formula	Description	State	$\Delta H_f^°$ kJ	$\Delta G_f^°$ kJ	$S^°$ J/deg	Source
				-1307.8		43
				-1309.6		86
				-1308.3		87
$H_3SiO_4^-$		aq		-1200.0		2
			-1426.2	-1253.9	112.5	6
			-1424.9	-1252.6	112.5	7
SiF_6^{2-}		aq	-2336.8	-2138.	-50.	1
			-2395.6	-2207.1	125.1	6
			-2389.1	-2199.4	122.2	31
			-2395.6	-2207.1	125.1	119
SiO		g	-99.6	-126.5	211.5	7
			-100.4	-127.3	211.6	21
			-99.6	-126.4	211.6	31
SiO_2	Quartz	c	-859.4	-805.0	41.8	3
			-910.9	-856.5	41.3	6
			-910.9	-856.7	41.8	7
			-910.6	-856.2	41.3	11
			-910.7	-856.3	41.5	21

Formula	Description	State	ΔH°_f kJ	ΔG°_f kJ	S° J/deg	Source
			-910.6	-856.2	41.3	22
			-910.9	-856.6	41.8	31
			-910.7	-856.3	41.4	98
			-910.7	-856.3	41.5	121
SiO_2	Cristobalite	c	-857.7	-803.7	42.6	3
			-908.3	-854.2	43.4	6
			-907.6	-853.9	43.4	7
			-908.0	-854.2	43.4	11
			-908.3	-854.5	43.4	21
			-906.9	-853.1	43.4	22
			-909.5	-855.4	42.7	31
			-907.9	-854.0	43.4	98
			-907.9	-854.5	44.9	108
SiO_2	Tridymite	c	-856.9	-802.9	43.3	3
			-909.0	-855.3	43.5	6
			-907.5	-853.9	43.9	7
					43.9	11
			-907.5	-853.8	43.9	21

Formula	Description	State	ΔH_f° kJ	ΔG_f° kJ	S° J/deg	Source
			−909.1	−855.3	43.5	31
SiO_2	Coesite	c	−905.9	−851.2	40.4	6
			−905.6	−850.9	40.4	7
			−904.2	−849.5	40.4	11
			−905.6	−850.8	40.4	21
			−906.3	−851.6	40.4	22
SiO_2	Stishovite	c	−861.5	−803.0	27.8	6
			−860.6	−801.9	27.8	11
			−861.3	−802.8	27.8	21
SiO_2		am	−847.3	−798.7	46.9	3
			−901.6	−849.0	46.9	6
				−850.3		7
			−903.2	−850.6	47.4	21
			−903.5	−850.7	46.9	31
			−901.6	−849.2	48.5	98
			−901.6	−848.6	46.9	119
$SiO_2 \cdot nH_2O$		am	−897.8	−848.9	60.0	22
SiO_2	Chalcedony	c	−909.1	−854.7	41.3	22

Formula	Description	State	ΔH°_f kJ	ΔG°_f kJ	S° J/deg	Source
$SiCl_4$		g	−609.6	−569.9	331.4	3
			−657.0	−617.0	330.6	6
			−657.0	−617.0	330.7	31
SiF_4		g	−1548.	−1506.	284.5	3
			−1614.9	−1572.6	282.2	6
			−1614.9	−1572.6	282.5	31
SiH_4		g	−61.9	−39.3	203.8	3
			34.3	56.9	204.6	31

- SILVER -

Formula	Description	State	ΔH°_f kJ	ΔG°_f kJ	S° J/deg	Source
Ag	Metal	c	0	0	42.7	3
			0	0	42.7	6
			0	0	42.7	11
			0	0	42.6	21
			0	0	42.6	22
			0	0	42.6	31
Ag^+		aq	105.9	77.1	73.9	3

Formula	Description	State	ΔH°_f kJ	ΔG°_f kJ	S° J/deg	Source
			105.6	77.1	72.9	6
			105.8	77.1	73.4	21
			105.6	77.1	72.7	31
			105.8	77.1		50
			105.6	77.1	72.9	119
Ag^{2+}		aq		268.2		1
			268.6	269.0	-88.	6
	unspc. aq. soln.		268.6	269.0	-88.	31
			268.6	269.0	-87.8	119
AgO^+		aq		225.5		1
AgO^-		aq		-23.0		1
$AgCl$		aq	-72.9	-73.0	154.8	6
			-72.8	-72.8	154.0	31
			-73.8	-72.9		50
			-72.9	-73.0	154.8	119
$AgCl_2^-$		aq	-245.0	-215.4	231.4	6
			-245.2	-215.4	231.4	31
			-258.6	-215.1		50

Formula	Description	State	ΔH_f° kJ	ΔG_f° kJ	S° J/deg	Source
			−245.0	−215.4	231.4	119
$AgCl_3^{2-}$		aq	−443.1	−345.9		50
$Ag(SO_3)_2^{3-}$		aq		−943.1		1
$Ag(S_2O_3)_2^{3-}$		aq	−1194.5	−1036.0		1
			−1278.7	−1027.2	73.6	6
			−1285.7			31
$Ag(NH_3)_2^+$		aq	−111.8	−17.4	241.8	1
			−111.3	−17.1	245.2	31
$Ag(CN)_2^-$		aq	269.9	301.4	205.0	3
			270.3	305.5	192.	31
Ag_2O		c	−30.6	−10.8	121.7	3
			−31.0	−11.2	121.3	6
			−31.0	−11.2	121.3	31
AgO		c	−25.1	10.9		1
Ag_2O_2		c	−24.3	27.6	117.	31
Ag_2O_3		c		87.0		1
			33.9	121.4	100.	31
$AgOH$		c		−92.0		12

Formula	Description	State	ΔH°_f kJ	ΔG°_f kJ	S° J/deg	Source
AgCl	Chlorargyrite	c	−127.0	−109.7	96.1	3
			−127.1	−109.8	96.2	6
			−127.1	−109.8	96.2	21
			−127.1	−109.8	96.2	31
			−127.1	−109.8		50
AgBr	Bromargyrite	c	−99.5	−95.9	107.1	3
			−100.8	−97.3	107.1	6
			−100.6	−97.1	107.1	21
			−100.4	−96.9	107.1	31
AgI	Iodargyrite	c	−62.4	−66.3	114.2	3
			−61.8	−66.2	115.5	6
			−61.8	−66.2	115.5	21
			−61.8	−66.2	115.5	31
Ag_2S α	Acanthite, orthorhombic	c	−31.8	−40.2	145.6	3
			−32.6	−40.7	144.0	6
			−31.6	−39.5	143.5	22
			−32.6	−40.7	144.0	31
			−29.7	−40.2		50

Formula	Description	State	ΔH_f° kJ	ΔG_f° kJ	S° J/deg	Source
Ag_2S β	Argentite	c	−29.3	−39.2	150.2	3
			−29.4	−39.5	150.6	6
			−29.4	−39.5	150.6	31
Ag_2S	Acanthite (argentite)	c	−32.3	−40.1	142.8	21
Ag_2CO_3		c	−506.1	−437.1	167.4	3
			−505.8	−436.9	167.4	6
			−505.8	−436.8	167.4	31
Ag_2SO_4		c	−713.4	−615.6	200.0	3
			−715.5	−617.8	199.8	6
			−715.9	−618.4	200.4	31
Ag_2SeO_4		c	−396.2	−286.6	181.2	3
			−431.4	−338.1	225.1	6
			−420.5	−334.2	248.5	31
$AgNO_3$		c	−123.1	−32.2	140.9	3
			−124.4	−33.4	140.9	6
			−124.4	−33.4	140.9	31
$AgMoO_4$		c		−821.7		1
Ag_2MoO_4		c	−837.8	−750.0	228.9	6

Formula	Description	State	ΔH_f° kJ	ΔG_f° kJ	S° J/deg	Source
			−840.6	−748.0	213.	31
Ag_2WO_4		c		−861.9		1
			−923.7	−841.4	252.7	6
			−925.5			31
Ag_2CrO_4		c	−737.2	−647.3	216.7	1
			−724.7	−634.8	217.6	6

– SODIUM –

Formula	Description	State	ΔH_f° kJ	ΔG_f° kJ	S° J/deg	Source
Na	Metal	c	0	0	51.0	3
			0	0	51.3	6
			0	0	51.2	11
			0	0	51.3	21
			0	0	51.2	31
Na^+		aq	−239.6	−261.9	60.2	3
			−240.4	−262.2	59.0	6
			−240.3	−261.9	58.4	21
			−240.1	−261.9	59.0	31

Sodium

Formula	Description	State	ΔH°_f kJ	ΔG°_f kJ	S° J/deg	Source
			−240.4			119
$NaCO_3^-$		aq		−797.2		2
			−935.9	−792.8	−49.8	31
$NaHCO_3^o$		aq	−930.9	−848.9	155.2	1
				−847.5		2
			−943.9	−849.7	113.8	31
$NaSO_4^-$		aq		−1008.0		2
			−1144.7	−1010.6	108.8	31
Na_2O		c	−415.9	−376.6	72.8	3
			−418.0	−379.1	75.1	6
			−415.9	−376.6	73.2	11
			−414.8	−376.1	75.3	21
			−414.8	−376.1	75.0	22
			−414.2	−375.5	75.1	31
NaOH		c	−426.7	−377.0	52.3	1
			−425.6	−379.4	64.4	6
			−425.8	−379.6	64.4	21
			−425.6	−379.5	64.4	31

Formula	Description	State	ΔH_f° kJ	ΔG_f° kJ	S° J/deg	Source
NaOH·H$_2$O		c	−732.9	−623.4	84.5	3
			−734.5	−629.3	99.5	31
NaCl	Halite	c	−411.0	−384.0	72.4	3
			−411.5	−384.5	72.1	6
			−411.3	−384.2	72.1	21
			−411.1	−384.1	72.1	22
			−411.1	−384.1	72.1	31
NaF	Villiaumite	c	−569.0	−541.0	58.6	3
			−575.2	−546.5	51.3	6
			−576.6	−546.3	51.3	21
			−573.6	−543.5	51.5	31
Na$_3$AlF$_6$	Cryolite	c	−3317.1	−3152.6	238.5	6
			−3309.5	−3144.9	238.4	21
			−3301.2	−3136.6	238.5	31
NaBr		c	−359.9	−347.7	85.8	1
			−361.4	−349.3	86.8	6
			−361.1	−349.0	86.8	31
NaI		c	−288.0	−237.2	92.5	1

Formula	Description	State	ΔH_f° kJ	ΔG_f° kJ	S° J/deg	Source
			−290.6	−287.4	98.5	6
			−287.8	−286.1	98.5	31
Na_2S		c	−373.2	−362.3	97.1	1
			−364.8	−349.8	83.7	31
Na_2CO_3		c	−1130.9	−1047.7	136.0	3
			−1131.4	−1047.8	135.0	6
			−1130.7	−1044.4	135.0	31
$Na_2CO_3 \cdot H_2O$	Thermonatrite	c	−1432.0	−1288.7	168.2	6
			−1431.3	−1285.3	168.1	31
				−1286.0		72
				−1286.6		88
$Na_2CO_3 \cdot 7H_2O$		c	−3200.0	−2714.2	422.2	31
				−2715.9		88
$Na_2CO_3 \cdot 10H_2O$	Natron	c	−4082.0	−3431.5	564.7	6
			−4081.3	−3427.7	562.7	31
				−3429.0		88
$NaHCO_3$	Nahcolite	c	−947.7	−851.9	102.1	3
			−913.4	−815.9	102.1	6

Formula	Description	State	ΔH°_f kJ	ΔG°_f kJ	S° J/deg	Source
			-950.8	-851.0	101.7	31
$Na_3(CO_3)(HCO_3) \cdot 2H_2O$	Trona	c		-2386.6		2
			-2684.9	-2383.4	301.2	31
				-2380.5		72
$NaAlCO_3(OH)_2$	Dawsonite	c	-1964.0	-1786.0	132.0	21
Na_2SO_4	Thenardite	c	-1384.5	-1266.8	149.5	3
			-1387.9	-1270.2	149.5	6
			-1387.8	-1270.0	149.6	21
			-1387.1	-1270.2	149.6	31
$Na_2SO_4 \cdot 10H_2O$	Mirabilite	c	-4327.9	-3647.2	585.8	6
			-4327.2	-3646.5	591.9	21
			-4327.3	-3646.8	592.0	31
$NaNO_3$	Soda-niter	c	-466.7	-365.9	116.3	3
			-468.3	-367.6	116.5	6
			-468.0	-367.2	116.5	21
			-467.8	-367.0	116.5	31
$Na_2B_4O_7 \cdot 4H_2O$	Kernite	c	-4507.4			31
$Na_2B_4O_7 \cdot 10H_2O$	Borax	c	-6278.9			21

Formula	Description	State	ΔH_f° kJ	ΔG_f° kJ	S° J/deg	Source
			−6288.6	−5516.0	586.	31
Na_2UO_4		c	−2096.			1
Na_2UO_4 α		c	−1887.0	−1768.6	166.0	21
			−1893.3	−1777.7	166.0	31
Na_3UO_4		c	−2021.5	−1897.4	198.2	21
			−2025.1	−1901.1	198.2	31
Na_2SiO_3		c	−1519.	−1427.	113.8	3
			−1557.6	−1463.7	113.8	6
			−1569.8	−1477.8	113.8	11
			−1554.9	−1462.8	113.8	31
			−1556.7	−1469.7	113.8	119
$NaSi_7O_{13}(OH)_3 \cdot 3H_2O$	Magadiite	c		−7373.0		69
$NaSi_{11}O_{20.5}(OH)_4 \cdot 3H_2O$	Kenyaite	c		−10893.0		69
$NaAlSiO_4$	Nepheline	c	−2075.3	−1960.7	124.3	6
			−2073.8	−1963.1	124.3	11
			−2092.1	−1977.5	124.4	21
			−2093.0	−1978.5	124.3	22
			−2092.8	−1978.1	124.3	31

Formula	Description	State	ΔH_f° kJ	ΔG_f° kJ	S° J/deg	Source
			−2054.9	−1939.8	124.3	36
			−2110.3	−1995.7	124.3	47
					123.8	53
$Na_{0.78}K_{0.22}AlSiO_4$	Nepheline	c	−2097.0	−1982.4	126.4	6
			−2110.3			21
$NaAlSi_3O_8$	Low albite	c	−3918.4	−3695.9	210.0	11
			−3935.1	−3711.7	207.4	21
			−3931.6	−3708.3	207.1	22
			−3935.1	−3711.5	207.4	31
			−3935.1	−3711.7	207.4	47
					204.6	53
$NaAlSi_3O_8$	Albite	c	−3921.0	−3698.6	210.0	6
			−3931.6	−3708.3	207.1	22
			−3922.9	−3700.3	210.1	36
$NaAlSi_3O_8$ β	Albite	c	−3909.9			6
$NaAlSi_3O_8$	High albite	c	−3907.4	−3690.4	228.7	11
			−3920.6	−3700.8	218.8	22
					223.9	53

Formula	Description	State	ΔH°_f kJ	ΔG°_f kJ	S° J/deg	Source
$NaAlSi_3O_8$	Analbite	c	-3924.2	-3706.5	226.4	21
			-3925.8	-3708.1	226.4	31
			-3924.2	-3706.5	226.4	47
$NaAlSi_3O_8$		am	-3860.2			6
			-3857.5	-3651.0	263.4	11
			-3875.5	-3665.3	251.9	21
			-3875.2	-3665.1	251.9	31
			-3875.4	-3665.3	251.9	47
$NaAlSi_2O_6$	Jadeite	c	-3009.7	-2830.8	133.5	11
			-3029.4	-2850.8	133.5	21
			-3021.3	-2842.8	133.5	22
			-3030.9	-2852.1	133.5	31
$NaAlSi_2O_6$	Dehydrated analcime	c	-2974.8	-2808.8	175.3	6
					175.4	21
			-2990.2	-2824.2	175.3	22
			-2985.3	-2819.1	175.3	31
$NaAlSi_2O_6 \cdot H_2O$	Analcime (analcite)	c	-3291.1	-3072.7	234.3	6
			-3298.2	-3078.6	234.3	11

Formula	Description	State	ΔH_f° kJ	ΔG_f° kJ	S° J/deg	Source
			−3309.8	−3091.7	234.4	21
			−3306.2	−3088.2	234.3	22
			−3300.8	−3082.6	234.3	31
			−3309.8	−3091.7	234.4	47
				−3084.3		99
$NaAlSi_2O_6 \cdot 2H_2O$	Analcime	c	−3288.1	−3068.3	234.4	36
$Na_2Al_3Si_3O_{10} \cdot 2H_2O$	Natrolite	c	−5718.6	−5316.6	359.7	59
$NaAl_3Si_3O_{10}(OH)_2$	Paragonite	c			281.6	6
			−5943.5	−5562.2	274.5	11
			−5928.6	−5548.0	277.8	22
					283.2	53
				−5558.9		99
$Na_{0.33}Al_{2.33}Si_{3.67}O_{10}(OH)_2$	Montmorillonite	c	−5718.8	−5346.1	262.8	20
			−5724.4	−5346.1	244.1	120
$(Na_{0.27}Ca_{0.10}K_{0.02})(Al_{1.52}Fe^{3+}_{0.19}$ $Mg_{0.22})(Si_{3.94}Al_{0.06})O_{10}(OH)_2$	Clay Spur montmorillonite	c		−5222.5		46
$4NaAlSi_3O_8 \cdot CaAl_2Si_2O_8$	Oligoclase	c	−3977.3	−3754.3	208.5	6
$3NaAlSi_3O_8 \cdot 2CaAl_2Si_2O_8$	Andesine	c	−4041.7	−3818.3	207.0	6

Formula	Description	State	ΔH°_f kJ	ΔG°_f kJ	S° J/deg	Source
$2.5NaAlSi_3O_8 \cdot 2.5CaAl_2Si_2O_8$	Labradorite	c	-4072.3	-3848.9	206.3	6
$Na_{0.676}Ca_{0.657}Al_{1.99}Si_{3.01}O_{10} \cdot$ $2.647H_2O$	Mesolite	c	-5947.1	-5513.2	363	59
$Na_2Mg_3Al_2Si_8O_{22}(OH)_2$	Glaucophane	c	-12072.6	-11339.0	538.6	11
$NaCa_2Fe^{2+}_4Al_3Si_6O_{22}(OH)_2$	Ferropargasite	c	-11237.2	-10538.8	689.7	11
$NaCa_2Mg_4Al_3Si_6O_{22}(OH)_2$	Pargasite	c	-12662.1	-11926.5	587.3	11
			-12623.4	-11912.6	669.4	22

- STRONTIUM -

Formula	Description	State	ΔH°_f kJ	ΔG°_f kJ	S° J/deg	Source
Sr		c	0	0	54.4	3
			0	0	53.1	6
			0	0	55.4	21
			0	0	52.3	31
Sr^{2+}		aq	-545.5	-557.3	-39.3	3
			-556.5	-571.4	-27.2	6
			-545.8	-559.4	-33.0	21
			-545.8	-559.5	-32.6	31

Formula	Description	State	ΔH°_f kJ	ΔG°_f kJ	S° J/deg	Source
Sr(OH)$^+$		aq	-556.5	-571.4	-27.2	119
			-781.6	-733.4	-5.8	6
				-721.3		31
			-781.6	-733.4	-5.8	119
SrO		c	-590.4	-559.8	54.4	3
			-604.3	-574.1	54.4	6
			-590.5	-560.4	55.5	21
			-592.0	-561.9	54.4	31
SrO$_2$		c	-642.7	-582.	54.	1
			-633.5			31
Sr(OH)$_2$		c	-959.4	-869.4	88.	1
			-959.0			31
			-959.4	-869.4	87.9	119
SrCl$_2$		c	-828.4	-781.2	117.	3
			-839.3	-791.2	114.8	6
			-828.9	-781.1	114.8	31
SrF$_2$		c	-1214.6	-1162.3	89.5	1
			-1232.4	-1180.6	82.1	6

Formula	Description	State	ΔH_f° kJ	ΔG_f° kJ	S° J/deg	Source
			−1216.3	−1164.8	82.1	31
SrBr$_2$		c	−729.3	−708.5	135.6	6
			−717.6	−695.9	135.1	21
			−717.6	−697.1	135.1	31
SrS		c	−452.3	−407.5	71.	1
					68.2	6
			−472.4	−467.8	68.2	31
SrCO$_3$	Strontianite	c	−1218.4	−1137.6	97.1	3
			−1232.6	−1152.3	97.1	6
			−1218.7	−1137.6	97.1	21
			−1232.6	−1152.6	97.1	22
			−1220.1	−1140.1	97.1	31
SrSO$_4$	Celestite	c	−1444.7	−1334.3	121.8	3
			−1467.8	−1353.4	110.9	6
			−1453.2	−1341.0	118.0	21
			−1453.1	−1340.7	117.2	22
			−1453.1	−1340.9	117.	31
Sr$_3$(PO$_4$)$_2$		c	−4130.9	−3899.9	293.	1

Formula	Description	State	ΔH°_f kJ	ΔG°_f kJ	S° J/deg	Source
			−4122.9			31
SrHPO$_4$		c	−1804.6	−1672.3	130.5	1
			−1821.7	−1688.6	121.	31
Sr(NO$_3$)$_2$		c	−988.7	−790.4	194.6	6
			−978.2	−779.1	194.6	21
			−978.2	−780.0	194.6	31
SrWO$_4$		c	−1666.5	−1533.4	158.2	1
			−1656.4	−1549.3	137.6	6
			−1639.7	−1531.	138.	31
SrSiO$_3$		c	−1553.1	−1467.7	94.1	1
			−1646.0	−1561.6	96.6	6
			−1630.1	−1561.5	96.1	11
			−1633.9	−1549.7	96.7	31
			−1627.8			47
			−1632.1	−1567.5	94.1	119
Sr$_2$SiO$_4$		c	−2178.2	−2074.0	180.	1
			−2329.0	−2215.1	153.1	6
			−2328.7	−2215.2	152.6	11

Formula	Description	State	ΔH°_f kJ	ΔG°_f kJ	S° J/deg	Source
			-2304.5	-2191.1	153.1	31
			-2302.9			47
			-2301.2	-2227.5	150.6	119
Sr_3SiO_5		c	-3031.0	-2886.9	205.0	11
			-2970.2	-2886.9	205.0	119

- SULFUR -

Formula	Description	State	ΔH°_f kJ	ΔG°_f kJ	S° J/deg	Source
S	Rhombic	c	0	0	31.9	3
			0	0	31.8	6
			0	0	31.8	7
			0	0	31.8	21
			0	0	31.8	31
S	Monoclinic	c	0.3	0.1	32.6	3
			0.3	0.1	32.6	6
			0.3			31
S		g	222.8	182.3	167.7	3
			278.8	238.2	167.8	31

Formula	Description	State	ΔH_f° kJ	ΔG_f° kJ	S° J/deg	Source
S_2		g	124.9	80.0		2
			128.4	79.1	228.1	6
			129.0	80.0	228.1	7
			128.5	79.4	228.2	21
			128.4	79.3	228.1	22
			128.4	79.3	228.2	31
S_8		g	102.3	49.7	430.9	6
			101.2	48.8	430.2	7
			101.2	48.8	430.3	21
			102.3	49.6	431.0	31
S^{2-}		aq	33.0	85.8	-14.6	6
			33.0	85.8	-14.6	7
				91.9		12
			33.0	85.8	-15.0	21
			33.1	85.8	-14.6	31
			30.9	81.4		50
			33.0	85.8	-14.6	119
S_2^{2-}		aq	30.1	79.5	28.4	6

Formula	Description	State	ΔH°_f kJ	ΔG°_f kJ	S° J/deg	Source
			30.1	79.5	28.4	7
				82.6		12
			30.1	79.5	28.5	31
S_3^{2-}		aq	25.9	73.6	66.1	6
			25.9	73.6	66.1	7
				75.2		12
			25.9	73.7	66.1	31
S_4^{2-}		aq	23.0	69.0	103.3	6
			23.0	69.0	103.3	7
				69.5		12
			23.0	69.1	103.3	31
S_5^{2-}		aq	21.3	65.7	140.6	6
			21.3	65.7	140.6	7
				65.6		12
			21.3	65.7	140.6	31
SO_3^{2-}		aq	−635.5	−485.8	−29.	1
			−635.5	−486.6	−29.3	6
			−635.5	−486.6	−29.3	7

Formula	Description	State	ΔH_f° kJ	ΔG_f° kJ	S° J/deg	Source
			-635.6	-486.6	-29.0	21
			-635.5	-486.5	-29.	31
			-635.5	-486.6	-29.3	119
SO_4^{2-}		aq	-907.5	-742.0	17.2	3
			-909.3	-743.8	17.6	6
			-909.3	-744.6	20.0	21
			-909.3	-744.5	20.1	31
			-909.6	-744.5		50
			-909.3	-743.8	17.6	119
$S_2O_3^{2-}$		aq	-644.	-532.2	121.	3
			-652.3	-513.8	37.2	6
			-652.3	-513.8	37.2	7
			-648.5	-522.5	67.	31
			-652.3	-513.8	37.2	119
$S_2O_4^{2-}$		aq	-686.	-577.	238.	3
			-753.5	-600.4	92.0	7
			-753.5	-600.3	92.	31
$S_2O_5^{2-}$		aq	-971.	-791.	105.	1

Formula	Description	State	ΔH°_f kJ	ΔG°_f kJ	S° J/deg	Source
$S_2O_6^{2-}$		aq	-1173.2	-966.	126.	1
	unspc. aq. soln.		-1198.3			31
$S_2O_8^{2-}$		aq	-1356.9	-1096.	146.	1
			-1338.9	-1110.4	248.1	7
			-1344.7	-1114.9	244.3	31
$S_3O_6^{2-}$		aq	-1167.	-958.	138.	1
	unspc. aq. soln.		-1199.6			31
$S_4O_6^{2-}$		aq	-1213.	-1030.5	259.	3
			-1224.2	-1040.4	257.3	31
$S_5O_6^{2-}$		aq	-1176.	-956.0	167.	1
	unspc. aq. soln.		-1236.4			31
HS^-		aq	-17.6	12.6	61.1	3
			-17.6	12.0	62.8	6
			-17.6	12.0	62.8	7
			-17.0	12.1	62.8	21
			-17.6	12.1	62.8	31
			-17.6	12.0		50
			-17.6	11.7	62.8	119

Formula	Description	State	ΔH_f° kJ	ΔG_f° kJ	S° J/deg	Source
H_2S		aq	−39.3	−27.4	122.2	3
			−39.7	−27.9	121.3	6
			−39.7	−27.9	121.3	7
			−39.7	−27.8	121.	31
			−42.2	−28.1		50
			−39.7	−27.9	121.3	119
HSO_3^-		aq	−628.0	−527.3	132.4	3
			−626.2	−527.8	139.7	6
			−626.2	−527.8	139.7	7
			−626.2	−527.7	139.7	31
			−626.2	−527.8	139.7	119
HSO_4^-		aq	−885.8	−752.9	126.8	3
			−888.8	−755.1	124.3	6
			−888.8	−755.1	124.3	7
			−887.3	−755.9	131.8	31
			−887.7	−755.9		50
			−888.8	−755.1	124.3	119
$HS_2O_3^-$		aq	−624.7	−523.6	162.8	6

Formula	Description	State	ΔH°_f kJ	ΔG°_f kJ	S° J/deg	Source
			-624.7	-523.6	162.8	7
				-541.8		12
			-624.7	-523.6	162.8	119
$HS_2O_4^-$		aq		-591.6		12
				-614.5		31
H_2SO_3		aq	-608.8	-538.0	234.	1
			-608.8	-537.9	232.2	6
			-608.8	-537.9	232.2	7
			-608.8	-537.8	232.2	31
			-608.8	-537.9	232.2	119
$H_2S_2O_3$		aq	-604.4	-527.0	242.2	6
			-604.4	-527.0	242.2	7
				-543.5		12
			-604.4	-527.0	242.2	119
$H_2S_2O_4$		aq	-686.	-585.8		1
				-616.6		31
$H_2S_2O_8$		aq	-1356.9	-1096.	146	1
			-1338.9	-1110.4	248.1	7

Sulfur

Formula	Description	State	ΔH_f° kJ	ΔG_f° kJ	S° J/deg	Source
SO		g	79.6	53.5	221.9	3
			6.2	−19.8	222.0	31
SO_2		g	−296.9	−300.4	248.5	3
			−296.8	−300.2	248.1	6
			−296.8	−300.2	248.1	7
			−296.8	−300.2	248.2	21
			−296.8	−300.2	248.2	31
SO_3		g	−395.2	−370.4	256.2	3
			−395.7	−371.1	256.6	6
			−395.7	−371.1	256.6	7
			−395.7	−371.0	256.8	21
			−395.7	−371.1	256.8	31
S_2Cl_2		l	−60.2	−24.7	167.	1
			−59.4			31
SF_6		g	−1096.2	−992.	290.8	3
			−1220.8	−1116.8	291.0	6
			−1209.	−1105.3	291.8	31
H_2S		g	−20.1	−33.0	205.6	3

Formula	Description	State	ΔH°_f kJ	ΔG°_f kJ	S° J/deg	Source
			−20.6	−33.6	205.7	6
			−20.6	−33.6	205.7	7
			−20.6	−33.5	205.8	21
			−20.6	−33.5	205.7	22
			−20.6	−33.6	205.8	31
H_2SO_4		l	−814.0	−690.0	156.9	21
			−814.0	−690.0	156.9	31
SO_2Cl_2		g		−307.9		1
			−364.0	−320.0	311.9	31

- TANTALUM -

Formula	Description	State	ΔH°_f kJ	ΔG°_f kJ	S° J/deg	Source
Ta	Metal	c	0	0	41.4	3
			0	0	41.5	6
			0	0	41.6	11
			0	0	41.5	21
			0	0	41.5	31
Ta_2O_5		c	−2091.6	−1969.0	143.1	3

Formula	Description	State	ΔH_f° kJ	ΔG_f° kJ	S° J/deg	Source
			-2047.2	-1912.5	143.1	6
			-2046.0	-1911.0	143.1	21
			-2046.0	-1911.2	143.1	31

- TELLURIUM -

Formula	Description	State	ΔH_f° kJ	ΔG_f° kJ	S° J/deg	Source
Te		c	0	0	49.7	3
			0	0	49.5	6
			0	0	49.5	21
			0	0	49.7	31
Te_2		g	171.5	121.3	268.1	3
			167.8	117.2	268.1	6
			168.2	118.0	268.1	31
Te^{2-}		aq		220.5		1
			101.2	174.0	-64.0	6
			101.2	174.0	-64.0	119
Te_2^{2-}		aq		162.1		1
				163.0		6

Tellurium

Formula	Description	State	ΔH°_f kJ	ΔG°_f kJ	S° J/deg	Source
Te^{4+}		aq		219.2		12
TeO_3^{2-}		aq	−532.6	−391.2	13.4	6
				−392.4		12
	unspc. aq. soln.		−544.8			31
			−532.6	−391.2	13.4	119
TeO_4^{2-}		aq		−456.4		12
$TeOOH^+$		aq		−258.5		1
				−256.9		6
H_2Te		aq		142.7		1
			77.8	89.5	141.0	6
			77.8	89.5	141.0	119
$HTeO_2^+$		aq		−261.5		12
$HTeO_3^-$		aq	−556.5	−452.3	138.9	6
				−436.6		12
			−556.5	−452.3	138.9	119
$HTeO_4^-$		aq		−515.8		12
H_2TeO_4		aq		−550.9		12
$TeCl_6^{2-}$		aq		−574.9		1

Formula	Description	State	ΔH°_f kJ	ΔG°_f kJ	S° J/deg	Source
TeO_2	Tellurite	c	-321.7	-264.6	58.6	6
				-273.3		12
			-322.6	-270.4	79.5	21
			-322.6	-270.3	79.5	31
H_2Te		g	154.4	138.5	234.	3
			99.7	-85.1	228.8	6
			99.6			31
H_2TeO_3		c	-613.0			6
				-478.5		12
H_6TeO_6		c		-1025.2		12
			-1298.7			31
TeF_6		g	-1318.0	-1222.	337.5	3
			-1369.0	-1273.0	335.8	6
			-1318.			31

Formula	Description	State	ΔH°_f kJ	ΔG°_f kJ	S° J/deg	Source

- THORIUM -

Formula	Description	State	ΔH°_f kJ	ΔG°_f kJ	S° J/deg	Source
Th	Metal	c	0	0	56.9	3
			0	0	53.4	6
			0	0	53.4	21
			0	0	53.4	31
Th^{4+}		aq	−765.7	−733.0	−314.	1
			−760.2	−723.8	−329.7	6
			−769.0	−705.1	−422.6	31
ThO_2	Thorianite	c	−1222.	−1164.8	70.7	1
			−1226.7	−1169.0	65.2	6
			−1226.4	−1168.8	65.2	21
			−1226.4	−1168.8	65.2	31
			−1226.7		65.3	119
$Th(OH)_4$	"Soluble"	c	−1763.6	−1586.	134.	1
$Th(OH)_4$		c	−1772.3	−1599.1	143.5	6
Th_2S_3		c	−1096.2	−1078.2	149.4	1
			−1084.	−1077.	180.	31

Formula	Description	State	ΔH_f° kJ	ΔG_f° kJ	S° J/deg	Source

- TIN -

Formula	Description	State	ΔH_f° kJ	ΔG_f° kJ	S° J/deg	Source
Sn	White	c	0	0	51.5	3
			0	0	51.5	6
			0	0	51.2	21
			0	0	51.6	31
			0	0	51.2	80
Sn	Gray	c	2.5	4.6	44.8	3
			-2.1	0.1	44.1	6
			-2.1	0.1	44.1	31
Sn^{2+}		aq	-10.0	-26.2	-24.7	1
			-24.4	-27.1	-69.9	6
			-8.8	-27.2	-17.	31
			-8.9	-27.2	-15.9	80
			-24.4	-27.1	-69.9	119
Sn^{4+}		aq		2.7		1
			30.5	2.5	-117.	31
SnO_3^{2-}		aq		-575.0		12

Formula	Description	State	ΔH°_f kJ	ΔG°_f kJ	S° J/deg	Source
$Sn_2O_3^{2-}$		aq		-590.3		12
$Sn(OH)^+$		aq		-252.3		6
				-253.6		12
			-286.2	-254.8	50.	31
				-252.3		119
$Sn(OH)_6^{2-}$		aq		-1299.1		1
$HSnO_2^-$		aq		-410.		1
SnF_6^{2-}		aq	-1986.1	-1757.	0	1
SnO	Romarchite	c	-286.2	-257.3	56.5	3
			-285.8	-256.9	56.5	6
			-285.8	-256.9	56.5	31
			-286.2	-257.3	56.5	80
SnO_2	Cassiterite	c	-580.8	-519.9	52.3	6
					52.3	11
				-515.5		12
			-580.7	-519.9	52.3	21
			-580.7	-519.6	52.3	31
			-580.7	-519.9	52.3	80

Formula	Description	State	ΔH°_f kJ	ΔG°_f kJ	S° J/deg	Source
$Sn(OH)_2$		c	−578.6	−492.0	96.6	3
$Sn(OH)_2$	Pptd.	c	−561.1	−491.6	155.	6
			−561.1	−491.6	155.	31
$Sn(OH)_4$		c	−1131.8	−951.9	121.	1
$Sn(OH)_4$	Pptd.	c	−1110.0			31
$SnCl_2$		c	−349.8	−302.0	122.6	1
			−325.1			31
			−325.1	−302.1	122.6	119
SnS	Herzenbergite	c	−77.8	−82.4	98.7	3
			−100.4	−98.5	77.0	6
			−106.5	−104.7	76.8	21
			−100.	−98.3	77.0	31
			−106.5	−104.7	76.8	80
SnS_2	Berndtite	c			87.4	6
					87.4	21
					87.4	31
			−153.6	−145.3	87.4	80
SnH_4		g		414.		12

Formula	Description	State	ΔH°_f kJ	ΔG°_f kJ	S° J/deg	Source
			162.8	188.3	227.7	31
$Sn(SO_4)_2$		c	-1646.0	-1451.0	155.2	1
				-1441.8		24
			-1629.2			31

- TITANIUM -

Formula	Description	State	ΔH°_f kJ	ΔG°_f kJ	S° J/deg	Source
Ti	Metal	c	0	0	30.3	3
			0	0	30.5	6
			0	0	30.6	21
			0	0	30.6	31
Ti^{2+}		aq		-314.2		1
Ti^{3+}		aq		-349.8		1
TiO^{2+}		aq		-577.		1
TiO_2^{2+}		aq		-467.2		12
$HTiO_3^-$		aq		-955.9		12
TiO		c	-519.4	-490.2	34.8	6
				-489.2		12

Formula	Description	State	ΔH_f° kJ	ΔG_f° kJ	S° J/deg	Source
			−542.7	−513.3	34.8	21
	α		−519.7	−495.0	50.	31
			−518.4	−489.1	34.7	119
TiO_2	Rutile	c	−944.5	−889.3	50.4	6
			−943.6	−888.4	50.4	11
				−888.4		12
			−944.8	−889.4	50.3	21
			−944.7	−889.5	50.3	31
			−946.0	−890.7	50.3	80
			−943.5	−888.2	50.2	119
TiO_2	Anatase	c	−938.1	−882.7	49.9	6
			−938.4	−883.0	49.9	11
			−938.7	−883.3	49.9	21
			−939.7	−884.5	49.9	31
			−912.5	−857.1	49.9	119
TiO_2	Brookite	c	−941.8			31
TiO_2	Hydrated	c	−866.	−821.3		1
	Hydrated, pptd.		−919.6			31

Formula	Description	State	ΔH_f° kJ	ΔG_f° kJ	S° J/deg	Source
Ti_2O_3		c	−1520.5	−1434.1	78.8	6
				−1432.2		12
			−1520.9	−1433.9	77.2	21
			−1520.9	−1434.2	78.8	31
			−1518.4	−1431.9	78.6	119
Ti_3O_5		c	−2458.5	−2316.9	129.4	6
				−2314.2		12
			−2459.2	−2317.4	129.4	21
			−2459.4	−2317.4	129.3	31
			−2458.7	−2317.0	129.3	119
Ti_4O_7		c	−3404.5	−3213.2	198.7	21
$TiO(OH)_2$		c		−1058.6		1
$Ti(OH)_3$		c		−1049.8		12
$TiCl_3$		c	−719.6	−664.8	139.7	6
			−721.7	−654.5	139.8	21
			−720.9	−653.5	139.7	31
			−690.4	−619.2	127.6	119
$TiBr_3$		c			176.6	6

Formula	Description	State	ΔH°_f kJ	ΔG°_f kJ	S° J/deg	Source
			−595.4	−570.7	176.4	21
			−548.5	−523.8	176.6	31

— TUNGSTEN —

W	Metal	c	0	0	33.5	3
			0	0	32.6	6
			0	0	32.6	21
			0	0	32.6	31
WO_4^{2-}		aq	−1115.4	−920.	63.	1
			−1073.4	−931.4	97.5	6
			−1075.7			31
			−1075.7	−679.7	40.6	80
			−1073.4	−931.4	97.5	119
WO_2		c	−570.3	−520.5	71.	1
			−589.7	−533.9	50.5	6
			−589.7	−533.8	50.5	21
			−589.7	−533.9	50.5	31

Formula	Description	State	ΔH°_f kJ	ΔG°_f kJ	S° J/deg	Source
WO_3	Yellow	c	−840.3	−763.4	83.3	3
WO_3 α		c	−842.9	−764.1	75.9	6
WO_3		c	−842.9	−764.1	75.9	21
			−842.9	−764.0	75.9	31
W_2O_5		c	−1413.8	−1284.1	142.	1
WS_2	Tungstenite	c	−193.7	−193.3	96.	3
			−202.9	−203.8	68.2	6
			−298.3	−297.9	95.0	21
			−209.			31

- URANIUM -

Formula	Description	State	ΔH°_f kJ	ΔG°_f kJ	S° J/deg	Source
U	Metal	c	0	0	50.3	3
			0	0	50.2	6
			0	0	50.3	21
			0	0	50.2	31
U^{3+}		aq	−514.6	−520.5	−126.	3
			−491.2	−489.1	−152.7	6

Formula	Description	State	ΔH°_f kJ	ΔG°_f kJ	S° J/deg	Source
			−514.6	−520.5	−126.0	21
			−489.1	−475.4	−192.	31
			−489.1	−480.7	−174.9	111
			−491.2	−489.1	−152.7	119
U^{4+}		aq	−613.8	−579.1	−326.	3
			−590.4	−550.2	−345.6	6
			−613.8	−579.1	−326.0	21
	Unhydrolyzed		−591.2	−531.0	−410.	31
			−591.2	−530.9	−414.	111
			−590.4	−583.7	−345.6	119
UO_2^+		aq	−1035.1	−994.1	50.	3
				−967.3		6
				−962.7		31
			−1032.6	−968.6	−25.	111
				−967.3		119
UO_2^{2+}		aq	−1047.7	−989.1	−71.	3
			−1024.2	−961.4	−86.2	6
			−1019.6	−953.5	−97.5	31

Formula	Description	State	ΔH°_f kJ	ΔG°_f kJ	S° J/deg	Source
			-1018.8	-952.7	-97.1	111
			-1024.2	-961.4	-86.2	119
$U(OH)^{3+}$		aq	-854.0	-809.6	-126.	3
			-827.2	-786.2	-115.5	6
			-828.1	-764.4	-192.	112
$(UO_2)_3(OH)_5^+$		aq		-3954.7		112
$UO_2CO_3^o$		aq	-1700.4	-1537.2	21.	115
$UO_2(CO_3)_2^{2-}$		aq		-2120.4		6
				-2104.1		115
$UO_2(CO_3)_3^{4-}$		aq		-2672.7		6
				-2677.8		70
				-2662.7		114
$UO_2(CO_3)_2(H_2O)_2^{2-}$		aq		-2602.4		70
UO_2SO_4		aq	-1955.2	-1730.9	-54.	3
			-1912.5	-1720.8	-54.0	6
			-1906.6	-1712.5	-54.	116
			-1912.5	-1720.8	-54.0	119
UO_2	Uraninite	c	-1084.9	-1032.1	78.2	6

Formula	Description	State	ΔH°_f kJ	ΔG°_f kJ	S° J/deg	Source
				-1031.8		12
			-1084.9	-1031.8	77.0	21
			-1084.9	-1031.7	77.0	31
				-1032.2		111
UO_3		c		-1142.2		12
UO_3 α		c	-1219.4			6
UO_3 β		c	-1221.7			6
	Orthorhombic, orange-red		-1220.3	-1142.3	96.3	31
UO_3 γ		c	-1225.3	-1148.0	98.6	6
			-1223.8	-1146.5	98.6	21
	Orthorhombic		-1223.8	-1145.9	96.1	31
$UO_3 \cdot H_2O$ α		c	-1528.0			31
$UO_3 \cdot H_2O$ β		c	-1533.9	-1394.8	126.	31
$UO_3 \cdot H_2O$ ε		c	-1531.3			31
$UO_3 \cdot 2H_2O$		c	-1826.7	-1630.8	167.	31
$U(OH)_3$		c		-1101.2		1
$U(OH)_4$		c		-1471.1		1
$UO_2(OH)_2$		c	-1534.7	-1400.8	141.8	6

Formula	Description	State	ΔH°_f kJ	ΔG°_f kJ	S° J/deg	Source
				-1349.9	-176.	112
				-1392.4		113
$UO_2(OH)_2 \cdot H_2O$	Schoepite	c	-1827.1	-1638.7	192.0	6
				-1828.4		70
				-1630.1	136.4	113
UCl_3		c	-891.2	-823.8	159.0	21
			-866.5	-799.1	159.0	31
UCl_4		c	-1048.9	-960.2	198.3	6
			-1018.4	-928.8	196.6	21
			-1019.2	-930.0	197.1	31
UF_4		c	-1924.6	-1833.8	151.7	6
			-1853.5	-1762.8	151.7	21
	Monoclinic		-1914.2	-1823.3	151.7	31
UO_2CO_3	Rutherfordine	c	-1696.2	-1570.2	146.4	6
			-1691.2	-1562.6	138.1	31
				-1577.4		70
			-1695.4	-1562.3	121.	115
$USiO_4$	Coffinite	c	-2000.	-1891.	118.8	117

Formula	Description	State	ΔH°_f kJ	ΔG°_f kJ	S° J/deg	Source

- VANADIUM -

Formula	Description	State	ΔH°_f kJ	ΔG°_f kJ	S° J/deg	Source
V	Metal	c	0	0	29.4	6
			0	0	28.9	21
			0	0	28.9	31
			0	0		35
V^{2+}		aq	−227.2	−217.1	−135.1	6
				−226.8		35
			−227.2	−217.1	−134.3	119
V^{3+}		aq	−263.2	−241.7	−238.5	6
				−251.4		35
			−263.2	−241.7	−238.5	119
VO^{2+}		aq	−488.7	−446.4	−140.6	6
			−486.6	−446.4	−133.9	31
				−456.0		35
VO_2^+		aq	−650.2	−587.2	−42.2	6
			−649.8	−587.0	−42.3	31
				−596.6		35

Formula	Description	State	ΔH_f° kJ	ΔG_f° kJ	S° J/deg	Source
$V_4O_9^{2-}$		aq		−2783.6		35
$V_{10}O_{28}^{6-}$		aq		−7792.3		35
$V(OH)^{2+}$		aq		−462.2		6
				−472.0		35
$HV_{10}O_{28}^{5-}$		aq		−7707		6
			−8694.	−7702.	222.	31
				−7846.2		35
$H_2V_{10}O_{28}^{4-}$		aq		−7728		6
				−7723.		31
				−7845.8		35
VO		c	−431.7	−404.1	38.9	6
			−431.8	−404.2	39.0	21
			−431.8	−404.2	38.9	31
V_2O_2		c		−790.8		35
V_2O_3	Karelianite	c	−1218.8	−1139.0	98.6	6
			−1218.8	−1139.0	98.1	21
			−1218.8	−1139.3	98.3	31
				−1133.9		35

Formula	Description	State	ΔH_f° kJ	ΔG_f° kJ	S° J/deg	Source
			-1252.7		98.3	119
V_2O_4		c	-1427.4	-1318.4	103.1	6
			-1427.2	-1318.4	103.5	21
			-1427.2	-1318.3	102.5	31
				-1330.5		35
V_2O_5		c	-1550.8	-1419.5	131.0	6
			-1550.6	-1419.4	130.5	21
			-1550.6	-1419.5	131.0	31
			-1592.4		131.0	119
V_2O_5	Aged ppt.	c		-1439.3		35
V_2O_5	Fresh ppt.	c		-1430.9		35
$VO(OH)_2$		c		-887.4		6
	Ppt.			-893.7		35
$V(OH)_3$	Ppt.	c		-912.1		35
VCl_2		c	-443.5	-397.1	97.1	6
			-451.9	-405.7	97.1	21
			-452.	-406.	97.1	31
VCl_3		c	-581.2	-511.7	131.0	6

Formula	Description	State	ΔH_f° kJ	ΔG_f° kJ	S° J/deg	Source
			−580.7	−511.4	131.0	21
			−580.7	−511.2	131.0	31

− YTTERBIUM −

Formula	Description	State	ΔH_f° kJ	ΔG_f° kJ	S° J/deg	Source
Yb		c	0	0	60.4	1
			0	0	59.9	6
			0	0	59.8	21
			0	0	59.9	31
Yb^{2+}		aq		−539.7		3
				−527.		31
Yb^{3+}		aq	−672.0	−656.0	−190.0	1
			−674.5	−644.0	−238.	31
Yb_2O_3		c	−1814.5	−1726.8	133.0	6
			−1814.6	−1726.8	133.0	21
			−1814.6	−1726.7	133.1	31
$Yb(OH)_3$		c		−1262.3		1
				−1298.3		6

Formula	Description	State	ΔH_f° kJ	ΔG_f° kJ	S° J/deg	Source
		- YTTRIUM -				
Y	Metal	c	0	0	47.3	1
			0	0	44.4	6
			0	0	44.4	21
			0	0	44.4	31
Y^{3+}		aq	-702.9	-686.6	-201.	1
			-701.6	-686.6	-202.1	6
			-723.4	-693.8	-251.	31
Y_2O_3		c	-1905.6	-1816.9	99.1	6
				-1682.0		12
			-1905.3	-1816.6	99.1	31
Y_2O_3	Cubic	c	-1905.3	-1816.6	99.1	21
$Y(OH)_3$		c	-1420.5	-1284.9	96.	1
			-1430.9	-1298.3	102.9	6
				-1291.1		31

Formula	Description	State	ΔH°_f kJ	ΔG°_f kJ	S° J/deg	Source
			- ZINC -			
Zn		c	0	0	41.6	3
			0	0	41.6	6
			0	0	41.6	21
			0	0	41.6	31
Zn^{2+}		aq	−152.4	−147.2	−106.5	3
			−153.7	−147.2	−110.7	6
			−153.4	−147.3	−109.6	21
			−153.9	−147.1	−112.1	31
			−153.7	−147.2	−110.7	119
ZnO_2^{2-}		aq		−389.2		1
				−384.2		31
$Zn(OH)^+$		aq		−333.2		6
				−329.3		12
				−330.1		31
				−333.2		119
$Zn(OH)_2$		aq		−535.4		6

Formula	Description	State	ΔH°_f kJ	ΔG°_f kJ	S° J/deg	Source
				-522.7		31
				-535.4		119
$Zn(OH)_3^-$		aq		-704.7		6
				-694.2		31
				-704.7		119
$Zn(OH)_4^{2-}$		aq		-871.3		6
				-858.5		31
				-871.3		119
$HZnO_2^-$		aq		-464.0		12
				-457.1		31
$Zn(NH_3)_4^{2+}$		aq		-307.5		1
			-533.5	-301.9	301.	31
$ZnCl^+$		aq		-275.3		31
$ZnCl_2$		aq		-403.7		31
$ZnCl_3^-$		aq		-540.5		31
$ZnCl_4^{2-}$		aq		-666.0		31
ZnO	Zincite	c	-350.7	-320.7	43.6	6
				-321.7		12

Formula	Description	State	ΔH_f° kJ	ΔG_f° kJ	S° J/deg	Source
			−350.5	−320.5	43.6	21
			−348.3	−318.3	43.6	31
			−351.4	−321.7	43.9	37
			−350.5	−320.5		50
$Zn(OH)_2$ α		c		−552.0		12
$Zn(OH)_2$ β		c		−554.5		6
				−557.0		12
			−641.9	−553.3	81.2	31
$Zn(OH)_2$ ε		c		−555.8		6
				−559.1		12
			−643.2	−555.1	81.6	31
$Zn(OH)_2$ γ		c		−557.8		12
				−553.8		31
$Zn(OH)_2$		am		−550.4		6
				−551.7		12
$ZnCl_2$		c	−415.9	−369.2	108.4	3
			−415.9	−369.4	108.8	6
			−415.0	−369.4	111.5	31

Formula	Description	State	ΔH°_f kJ	ΔG°_f kJ	S° J/deg	Source
ZnBr$_2$		c	-327.1	-310.2	137.4	3
			-329.7	-312.6	136.4	6
			-328.6	-312.1	138.5	31
ZnS	Sphalerite	c	-202.9	-198.3	57.7	3
			-208.4	-203.8	57.7	6
			-206.9	-202.5	58.7	21
			-205.0	-200.6	58.6	22
			-206.0	-201.3	57.7	31
			-206.9	-202.5		50
ZnS	Pptd.	c	-185.4	-180.7		1
ZnS	Wurtzite	c	-189.5	-184.9	57.7	1
			-195.0	-193.3	67.8	6
			-194.6	-190.2	58.8	21
			-191.8	-187.5	58.8	22
			-192.6			31
ZnSe	Stilleite	c	-142.	-145.2	93.3	1
			-164.4	-168.2	82.8	6
			-163.	-163.	84.	31

Formula	Description	State	ΔH_f° kJ	ΔG_f° kJ	S° J/deg	Source
$ZnCO_3$	Smithsonite	c	-812.5	-731.4	82.4	3
			-815.0	-733.9	82.4	6
			-812.8	-731.5	82.4	21
			-812.8	-731.6	82.4	22
			-812.8	-731.5	82.4	31
			-812.8	-731.5		50
$Zn_5(OH)_6(CO_3)_2$	Hydrozincite	c	-3465.4	-3141.2		50
$ZnSO_4$	Zinkosite	c	-978.6	-871.6	124.7	3
			-979.8	-868.6	110.4	6
			-982.8	-871.5	110.5	21
			-982.8	-871.5	110.5	31
$ZnSO_4 \cdot H_2O$		c	-1299.6	-1129.3	146.0	3
			-1304.5	-1132.0	138.5	31
$ZnSO_4 \cdot 6H_2O$	Bianchite	c	-2775.2	-2322.1	363.2	3
			-2776.8	-2319.9	363.6	6
			-2777.5	-2324.4	363.6	21
			-2777.5	-2324.4	363.6	31
$ZnSO_4 \cdot 7H_2O$	Goslarite	c	-3075.6	-2560.2	386.6	3

Formula	Description	State	ΔH°_f kJ	ΔG°_f kJ	S° J/deg	Source
			-3077.1	-2562.4	388.7	6
			-3077.8	-2562.6	388.7	21
			-3077.8	-2562.7	388.7	31
$ZnWO_4$	Sanmartinite	c	-1232.6	-1123.7	119.3	21
$ZnAl_2O_4$	Gahnite	c	-2065.2			31
$ZnSiO_3$		c	-1232.6	-1149.8	89.5	1
			-1284.0	-1201.0	89.5	11
			-1260.2			31
			-1284.3	-1201.4	89.5	119
Zn_2SiO_4	Willemite	c	-1641.6	-1528.1	131.4	6
			-1636.5	-1522.9	131.4	21
			-1636.7	-1523.2	131.4	31
			-1636.3	-1522.9	131.4	119
Zn_2TiO_4		c	-1651.0	-1537.6	143.1	6
			-1647.7	-1534.0	143.1	21
			-1647.7	-1534.2	143.1	31

Formula	Description	State	ΔH°_f kJ	ΔG°_f kJ	S° J/deg	Source
		– ZIRCONIUM –				
Zr		c	0	0	38.4	3
			0	0	38.9	6
			0	0	39.0	21
			0	0	39.0	31
Zr^{4+}		aq		-524.5		6
				-594.		12
ZrO^{2+}		aq		-843.1		12
$HZrO_3^-$		aq		-1203.7		1
ZrO_2	Baddeleyite	c	-1100.6	-1042.9	50.7	6
				-1036.4		12
			-1100.6	-1042.8	50.4	21
			-1100.6	-1042.8	50.4	31
			-1094.1	-1036.4	50.7	119
$Zr(OH)_4$		c	-1720.5	-1548.	130.	1
			-1661.5	-1487.9	127.6	6
			-1720.5	-1548.1	129.7	119

Formula	Description	State	ΔH°_f kJ	ΔG°_f kJ	S° J/deg	Source
$ZrO(OH)_2$		c	-1414.2	-1303.3	92.	1
$ZrSiO_4$	Zircon	c	-1990.3	-1876.1	84.5	6
			-2025.0	-1910.8	84.5	11
			-2033.4	-1918.9	84.0	21
			-2033.4	-1919.1	84.1	31
			-2022.1	-1911.9	84.5	119

REFERENCES
Alphabetical order

REFERENCES (alphabetical order)

27 Apps, J. A. (1970). The Stability Field of Analcime, unpub. Ph.D. Thesis, Harvard University, Cambridge, MA.

119 Babushkin, V. I., G. M. Matveyev and O. P. Mchedlov-Petrossyan (1985). Thermodynamics of Silicates, Springer-Verlag, Berlin, 459 pp.

112 Baes, C. F., Jr. and R. E. Mesmer (1976). The Hydrolysis of Cations, John Wiley and Sons, New York, 489 pp.

92 Barton, M. D. (1982). The thermodynamic properties of topaz solid solutions and some petrologic applications, Amer. Miner., $\underline{67}$, 956-974.

91 Barton, M. D., H. T. Haselton, Jr., B. S. Hemingway, O. J. Kleppa and R. A. Robie (1982). The thermodynamic properties of fluor-topaz, Amer. Miner., $\underline{67}$, 350-355.

37 Barton, P. B., Jr. and P. M. Bethke (1960). Thermodynamic properties of some synthetic zinc and copper minerals, Amer. J. Sci., $\underline{258-A}$, 21-34.

17 Beane, R. E. (1974). Biotite stability in the porphyry copper environment, Econ. Geol., $\underline{69}$, 241-256.

75 Bird, G. W. and G. M. Anderson (1973). The free energy of formation of magnesian cordierite and phlogopite, Amer. J. Sci., $\underline{273}$, 84-91.

56 Bricker, O. P. (1965). Some stability relationships in the system $Mn-O_2-H_2O$ at 25°C and 1 atm. total pressure, Amer. Miner., $\underline{50}$, 1296-1354.

69 Bricker, O. P. (1969). Stability constants and Gibbs free energies of formation of magadiite and kenyaite, Amer. Miner., $\underline{54}$, 1026-1033.

13 Bricker, O. P., H. W. Nesbitt and W. D. Gunter (1973). The stability of talc, Amer. Miner., $\underline{58}$, 64-72.

70 Bullwinkel, E. P. (1954). The chemistry of uranium in carbonate solutions, U. S. Atomic Energy Commission, Raw Materials Division RMO, 2614, 59 pp.

51 Charlu, T. V., R. C. Newton and O. J. Kleppa (1975). Enthalpies of formation at 970K of compounds in the system $MgO-Al_2O_3-SiO_2$ from high temperature solution calorimetry, Geochim. et Cosmochim. Acta, 39, 1487–1497.

84 Chernosky, J. V., Jr. (1974). The upper stability of clinochlore at low pressure and the free energy of formation of Mg-cordierite, Amer. Miner., 59, 496–507.

15 Christ, C. L., P. B. Hostetler and R. M. Siebert (1973). Studies in the system $MgO-SiO_2-CO_2-H_2O$ (III): The activity-product constant of sepiolite, Amer. J. Sci., 273, 65–83.

76 Christ, C. L., P. B. Hostetler and R. M. Siebert (1974). Stabilities of calcite and aragonite, J. Research U. S. Geol. Survey, 2, 175–184.

71 Couturier, Y., G. Michard and G. Sarazin (1984). Constantes de formation des complexes hydroxydes de l'aluminium en solution aqueuse de 20 a 70°C, Geochim. et Cosmochim. Acta, 48, 649–660.

100 Crerar, D. A. and H. L. Barnes (1974). Deposition of deep-sea manganese nodules, Geochim. et Cosmochim. Acta, 38, 279–300.

35 Evans, H. T., Jr., and R. M. Garrels (1958). Thermodynamic equilibria of vanadium in aqueous systems as applied to the interpretation of the Colorado Plateau ore deposits, Geochim. et Cosmochim. Acta, 15, 131–149.

19 Feitknecht, W. and P. W. Schindler (1963). Solubility constants of metal oxides, metal hydroxides and metal hydroxide salts in aqueous solutions, Pure Appl. Chem., 6, 130–199.

111 Fuger, J. and F. L. Oetting (1976). The chemical thermodynamics of actinide elements and compounds. Pt. 2, The actinide aqueous ions, Internat. Atomic Energy Agency, Vienna, 65 pp.

82 Garrels, R. M. (1957). Some free energy values from geologic relations, Amer. Miner., 42, 780–792.

26 Garrels, R. M. (1984). Montmorillonite/illite stability diagrams, Clays and Clay Miner., 32, 161–166.

2   Garrels, R. M. and C. L. Christ (1965). <u>Solutions, Minerals and Equi-libria</u>, Freeman, Cooper and Co., San Francisco, Calif., 450 pp.

65   Garrels, R. M., M. E. Thompson and R. Siever (1960). Stability of some carbonates at 25°C and 1 atmosphere total pressure, Amer. J. Sci., <u>258</u>, 402-418.

73   Haas, H. and M. J. Holdaway (1973). Equilibria in the system $Al_2O_3-SiO_2-H_2O$ involving the stability limits of pyrophyllite, and thermodynamic data of pyrophyllite, Amer. J. Sci., <u>273</u>, 449-464.

74   Haas, H. and M. J. Holdaway (1974). Equilibria in the system $Al_2O_3-SiO_2-H_2O$ involving the stability limits of pyrophyllite, and thermodynamic data of pyrophyllite: Additional data, Amer. J. Sci., <u>274</u>, 825-828.

72   Harvie, C. E., N. Moller and J. H. Weare (1984). The prediction of mineral solubilities in natural waters: The $Na-K-Mg-Ca-H-Cl-SO_4-OH-HCO_3-CO_3-CO_2-H_2O$ system to high ionic strengths at 25°C, Geochim. et Cosmochim. Acta, <u>48</u>, 723-751.

20   Helgeson, H. C. (1969). Thermodynamics of hydrothermal systems at elevated temperatures and pressures, Amer. J. Sci., <u>267</u>, 729-804.

80   Helgeson, H. C. (1983 and 1984). Supcrt Update Notices.

22   Helgeson, H. C., J. M. Delany, W. H. Nesbitt and D. K. Bird (1978). Summary and critique of the thermodynamic properties of rock-forming minerals, Amer. J. Sci., <u>278-A</u>, 229 pp.

32   Hem, J. D. (1978a). Redox processes at surfaces of manganese oxide and their effects on aqueous metal ions, Chem. Geol., <u>21</u>, 199-218.

33   Hem, J. D. (1978b). Redox co-precipitation mechanisms of manganese oxides, <u>In</u>: Kavanaugh, M. C. and J. O. Leckie, eds., <u>Particulates in Water</u>, Advances in Chemistry Series 189, Amer. Chem. Soc., Washington, D. C., 45-72.

57   Hem, J. D. and C. E. Roberson (1967). Form and stability of aluminum hydroxide complexes in dilute solution, U. S. Geol. Survey Water-Supply Paper 1827-A, A1-A55.

64   Hem, J. D., C. E. Roberson and C. J. Lind (1985). Thermodynamic stability of CoOOH and its coprecipitation with manganese, Geochim. et Cosmochim. Acta, <u>49</u>, 801-810.

121   Hemingway, B. S., J. L. Haas, Jr. and G. R. Robinson, Jr. (1982) Thermodynamic properties of selected minerals in the system $Al_2O_3-CaO-SiO_2-H_2O$ at 298.15 K and 1 bar (10^5 pascals) pressure and at higher temperatures, U. S. Geol. Survey Bull. 1544, 70 pp.

47   Hemingway, B. S. and R. A. Robie (1977a). Enthalpies of formation of low albite ($NaAlSi_3O_8$), gibbsite ($Al(OH)_3$) and $NaAlO_2$; revised values for $\Delta H^\circ_{f,298}$ and $\Delta G^\circ_{f,298}$ of some aluminosilicate minerals, J. Research U. S. Geol. Survey, <u>5</u>, 413-429.

52   Hemingway, B. S. and R. A. Robie (1977b). The entropy and Gibbs free energy of formation of the aluminum ion, Geochim. et Cosmochim. Acta, <u>41</u>, 1402-1404.

28   Hemingway, B. S., R. A. Robie and J. A. Kittrick (1978). Revised values for the Gibbs free energy of formation of $[Al(OH)_4^-$ aq], diaspore, boehmite and bayerite at 298.15K and 1 bar, the thermodynamic properties of kaolinite to 800K and 1 bar, and the heats of solution of several gibbsite samples, Geochim. et Cosmochim. Acta, <u>42</u>, 1533-1543.

67   Hemingway, B. S. and R. A. Robie (1984). Heat capacity and thermodynamic functions for gehlenite and staurolite: with comments on the Schottky anomaly in the heat capacity of staurolite, Amer. Miner., <u>69</u>, 307-318.

48   Hemley, J. J., P. B. Hostetler, A. J. Gude and W. T. Mountjoy (1969). Some stability relations of alunite, Econ. Geol., <u>64</u>, 599-612.

43   Hemley, J. J., J. W. Montoya, C. L. Christ and P. B. Hostetler (1977a). Mineral equilibria in the $MgO-SiO_2-H_2O$ system: I. Talc-chrysotile-forsterite-brucite stability relations, Amer. J. Sci., <u>277</u>, 322-356.

44   Hemley, J. J., J. W. Montoya, D. R. Shaw and R. W. Luce (1977b). Mineral equilibria in the $MgO-SiO_2-H_2O$ system: II. Talc-antigorite-forsterite-anthophyllite-enstatite stability relations and some geologic implications in the system, Amer. J. Sci., <u>277</u>, 353-383.

63 Hemley, J. J., J. W. Montoya, J. W. Marinenko and R. W. Luce (1980). Equilibria in the system Al_2O_3-SiO_2-H_2O and some general implications for alteration/mineralization processes, Econ. Geol., 75, 210-228.

85 Hepler, L. G. and G. Olofsson (1975). Mercury: Thermodynamic Properties, Chemical Equilibria, and Standard Potentials, Chem. Rev., 75, 585-602.

14 Hostetler, P. B. and C. L. Christ (1968). Studies in the system MgO-SiO_2-CO_2-H_2O (I): The activity-product constant of chrysotile, Geochim. et Cosmochim. Acta, 32, 485-497.

46 Huang, W. H. and W. D. Keller (1973). Gibbs free energies of formation calculated from dissolution data using specific mineral analyses. III. Clay minerals, Amer. Miner., 58, 1023-1028.

59 Johnson, G. K., H. E. Flotow and P. A. G. O'Hare (1983). Thermodynamic studies of zeolites: natrolite, mesolite and scolecite, Amer. Miner., 68, 1134-1145.

118 Johnson, G. K., H. E. Flotow, P. A. G. O'Hare and W. S. Wise (1985). Thermodynamic studies of zeolites: heulandite, Amer. Miner., 70, 1065-1071.

24 Karpov, I. K., S. A. Kashik and V. D. Pampura (1968). Thermodynamic Constants for Calculations in Geochemistry and Petrology (In Russian), Izdva, Nauka, 143 pp.

11 Karpov, I. K., A. I. Kiselev and F. A. Letnikov (1971). Chemical Thermodynamics in Petrology and Geochemistry, Akademia Nauka, Irkutsk, 385pp.

77 Karpov, I. K., A. I. Kiselev and F. A. Letnikov (1976). Modeling of natural mineral assemblages with an IBM, Nedra, Moscow, 256 pp.

81 Kelley, K. K. (1937). Contributions to the data on theoretical metallurgy. VII. The thermodynamic properties of sulfur and its inorganic compounds, U. S. Bureau Mines Bull. 406, 154 pp.

83 King, E. G., R. Barany, W. W. Weller and L. B. Pankratz (1967). Thermodynamic properties of forsterite and serpentine, U. S. Bureau Mines Rept. Inv. 6962, 19 pp.

90 Kittrick, J. A. (1966). The free energy of formation of gibbsite and Al(OH)$_4^-$ from solubility measurements, Soil Sci. Soc. Amer. Proc., **30**, 595-598.

86 Kittrick, J. A. (1970). Precipitation of kaolinite at 25°C and 1 atm., Clays and Clay Miner., **18**, 261-267.

38 Kittrick, J. A. (1971a). Stability of montmorillonites: I. Belle Fourche and Clay Spur montmorillonites, Soil Sci. Soc. Amer. Proc. **35**, 140-145.

39 Kittrick, J. A. (1971b). Stability of montmorillonites: II. Aberdeen montmorillonite, Soil Sci. Soc. Amer. Proc. **35**, 820-823.

117 Langmuir, D., and K. Applin (1977). Refinement of the thermodynamic properties of uranium minerals and dissolved species with applications to the chemistry of ground waters in sandstone-type uranium deposits, In: Campbell, J. A., ed., Short Papers of the U. S. Geological Survey Uranium-Thorium Symp., U. S. Geol. Survey Circ. **753**, 57-66.

94 Langmuir, D. and A. C. Riese (1985). The thermodynamic properties of radium, Geochim. et Cosmochim. Acta, **49**, 1593-1601.

1   Latimer, W. M. (1952). Oxidation Potentials, Second edition, Prentice-Hall, New York, 392 pp.

60 Leussing, D. L. and I. M. Kolthoff (1953). The solubility product of ferrous hydroxide and the ionization of the aquo-ferrous ion, J. Amer. Chem. Soc., **75**, 2746.

34 Lind, C. J. (1978). Polarographic determination of lead hydroxide formation constants at low ionic strength, Envir. Sci. Tech., **12**, 1406-1410.

25 Lippmann, F. (1979). Stabilitätsbeziehungen der Tonminerale, Neues Jharbuch. Miner., **136**, 287-309.

102 Mann, A. W. and R. L. Deutscher (1980). Solution geochemistry of lead and zinc in water containing carbonate, sulfate, and chloride ions, Chem. Geol., **29**, 293-311.

42 Mattigod, S. V. and G. Sposito (1978). Improved method for estimating the standard free energies of formation ($\Delta G^{\circ}_{f298.15}$) of smectites, Geochim. et Cosmochim. Acta, 42, 1753–1762.

23 May, H. M., P. A. Helmke and M. L. Jackson (1979). Gibbsite solubility and thermodynamic properties of hydroxy-aluminum ions in aqueous solutions at 25°C, Geochim. et Cosmochim. Acta, 43, 861–868.

7   Mel'nik, Y. P. (1972). Thermodynamic Constants for the Analysis of Conditions of Formation of Iron Ores (In Russian), Naukova Dumka, Kiev, 196 pp.

61 Morgan, J. J. (1967). Chemical equilibria and kinetic properties of manganese in natural waters, In: Faust, S. D. and J. V. Hunter, eds., Principles and Applications of Water Chemistry, John Wiley and Sons, New York, 561–624.

6   Naumov, G. B., B. N. Ryzhenko and I. L. Khodakovsky (1974). Handbook of Thermodynamic Data, Natl. Tech. Inf. Service, Pb-226, 722/7GA, U. S. Dept. Commerce, 328 pp.

121 Nesbitt, H. W. (1977). Estimation of the thermodynamic properties of Na- Ca- and Mg- beidellites, Canadian Miner., 15, 22–30.

107 Newberg, Donald W. (1967). Geochemical implications of chrysocolla- bearing alluvial gravels, Econ. Geol. 62, 932–956.

113 Nikitin, A. A., Z. I. Sergeyeva, I. L. Khodakovsky and G. B. Naumov (1972). Uranyl hydrolysis in a hydrothermal region, Geokhimiya, 3, 297–307.

114 O'Cinneide, S., J. P. Scanlan and M. J. Hynes (1972). Overall stability constant of the complex $UO_2(CO_3)_3^{-4}$, Chem. Indus. (London), 50, 340.

29 Parks, G. A. (1972). Free energies of formation and aqueous solubilities of aluminum hydroxides and oxide hydroxides at 25°C, Amer. Miner., 57, 1163–1189.

105 Perkins, D., III, E. J. Essene, E. F. Westrum, Jr. and V. J. Wall (1979). New thermodynamic data for diaspore and their applica- tion to the system $Al_2O_3-SiO_2-H_2O$, Amer. Miner., 64, 1080–1090.

87   Polzer, W. L. and J. D. Hem (1965). The dissolution of kaolinite,
     J. Geophys. Research, 70, 6233-6240.

12   Pourbaix, M. (1960). Standard free energies of formation at 25°C,
     CEBELCOR Rapt. 684, Belgian Center for the Study of Corrosion,
     Brussels, 57 pp.

62   Pourbaix, M. (1963). Atlas d'equilibres electrochimiques a 25°C,
     Gauthier-Villars, Paris, 644 pp.

93   Pourbaix, M. (1973). Lectures on Electrochemical Corrosion, Plenum
     Press, New York and London, 336 pp.

55   Pourbaix, M. and X. Yang (1981). Chemical and electrochemical
     equilibria in the presence of a gaseous phase. 5. -oxygen-
     hydrogen-iron, CEBELCOR Rapt. Tech. 1401, 260, 116 pp.

89   Raupach, M. (1963a). Solubility of simple aluminum compounds expected in
     soils. I. Hydroxides and oxyhydroxides, Australian J. Soil Research,
     1, 28-35.

68   Raymahashay, B. C. (1968). A geochemical study of rock alteration by hot
     springs in the Paint Pot Hill area, Yellowstone Park, Geochim. et
     Cosmochim. Acta, 32, 499-522.

49   Reeseman, A. L. (1974). Aqueous dissolution studies of illite under
     ambient conditions, Clays and Clay Miner., 22, 443-454.

78   Reeseman, A. L. and W. D. Keller (1968). Aqueous solubility studies of
     high-alumina and clay minerals, Amer. Miner., 53, 929-942.

95   Reinert, M. (1965). Über die Bestimmung der Löslichkeit schwerlöslicher
     Salze mit basischen Anionen, unpub. Ph. D. Thesis, University of Bern.

98   Richet, P., Y. Bottinga,, L. Denielou,, J. P. Petitet and C. Tequi
     (1982). Thermodynamic properties of quartz, cristobalite and amorphous
     SiO_2: drop calorimetry measurements between 1000 and 1800K and a
     review from 0 to 2000K, Geochim. et Cosmochim. Acta, 46, 2639-2658.

104 Rickard, D. T. (1970). The chemistry of copper in natural aqueous solutions, Acta Universitatis Stockholmiensis, Stockholm Contributions in Geology, $\underline{23}$:1, 64pp.

101 Rickard, D. T. and J. O. Nriagu (1978). Aqueous environmental chemistry of lead, In: Nriagu, J. O., ed., The Biogeochemistry of Lead in the Environment, Elsevier - North Holland Biomedical Press, New York, 219-284.

106 Robie, R. A., H. T. Haselton, Jr. and B. S. Hemingway (1984). Heat capacities and entropies of rhodochrosite ($MnCO_3$) and siderite ($FeCO_3$) between 5 and 600 K, Amer. Miner., $\underline{69}$, 349-357.

21   Robie, R. A., B. S. Hemingway and J. R. Fisher (1978). Thermodynamic properties of minerals and related substances at 298.15 K and 1 bar (10^5 Pascals) pressure and at higher temperatures, U. S. Geol. Survey Bull. 1452, 456 pp.

66   Robie, R. A. and B. S. Hemingway (1984). Entropies of kyanite, andalusite and sillimanite: additional constraints on the pressure and temperature of the Al_2SiO_5 triple point, Amer. Miner., $\underline{69}$, 298-306.

110 Robie, R. A., B. S. Hemingway, J. Ito and K. M Krupka (1984). Heat capacity and entropy of Ni_2SiO_4-olivine from 5 to 1000K and heat capacity of Co_2SiO_4 from 360 to 1000K, Amer. Miner., $\underline{69}$, 1096-1101.

30   Robie, R. A. and D. R. Waldbaum (1968). Thermodynamic properties of minerals and related substances at 298.15°K (25°C) and one atmosphere (1.013 bars) pressure and at higher temperatures, U. S. Geol. Survey Bull. 1259, 256 pp.

108 Robinson, G. R., Jr., J. L. Haas, Jr., C. M. Schafer and H. T. Haselton, Jr. (1982). Thermodynamic and thermophysical properties of selected phases in the $MgO-SiO_2-H_2O-CO_2$, $CaO-Al_2O_3-SiO_2-H_2O-CO_2$, and $Fe-FeO-Fe_2O_3-SiO_2$ chemical systems, with special emphasis on the properties of basalts and their mineral components, U. S. Geol. Survey Open-file Report 83-79. 429 pp.

3   Rossini, F. D., D. D. Wagman, W. H. Evans, S. Levine, and I. Jaffe
    (1952). <u>Selected Values of Chemical Thermodynamic Properties</u>, Nat'l.
    Bureau Standards Circ. 500, U. S. Dept. Commerce. Washington, D. C.

40 Routson, R. C. and J. A. Kittrick (1971). Illite solubility, Soil
    Sci. Soc. Amer. Proc., <u>35</u>, 714-718.

88 Saegusa, F. (1950). Science reports, Tohoku Univ., 1st Series, <u>34</u>, 104.

50 Sangameshwar, S. R. and H. L. Barnes (1983). Supergene processes in zinc-
    lead-silver sulfide ores in carbonates, Econ. Geol., <u>78</u>, 1379-1397.

53 Saxena, S. K. (1976). Entropy estimates for some silicates at 298°K
    from molar volumes, Science, <u>193</u>, 1241-1242.

96 Schindler, P., H. Althaus, F. Hofer and W. Minder (1965). Löslichkeits-
    produkte von Zinkoxid, Kupferhydroxid und Kupferoxid in Abhängigkeit
    von Teilehengrösse und molarer Oberfläche. Ein Beitrag zur Thermody-
    namik von Grenzflächen fest-flüssig. Helv. Chim. Acta., <u>48</u>, 1204-1215.

115 Sergeyeva, E. I., A. A. Nikitin, I. L. Khodakovsky and G. B. Naumov
    (1972). Experimental investigation in the system UO_3-CO_2-H_2O in
    25-200°C temperature interval, Geochemistry Internat., <u>9</u>, 900-910.

97 Sillen, L. G. and A. E. Martell (1964). <u>Stability Constants of Metal-Ion
    Complexes</u>, Spec. Publ. 17, The Chem. Soc., Burlington House, London,
    754 pp.

79 Silman, J. F. B. (1958). The stabilities of some oxidized copper
    minerals in aqueous solutions at 25°C and 1 atmosphere total pressure,
    unpub. Ph.D. thesis, Harvard University, Cambridge, MA, 98 pp.

9   Stull, D. R. and H. Prophet (1971). <u>JANAF Thermochemical Tables</u>, Second
    edition, Nat'l. Standard Reference Data Series, Nat'l. Bureau
    Standards, <u>37</u>.

103 Symes, J. L. and D. R. Kester (1984). Thermodynamic stability studies
    of the basic copper carbonate mineral, malachite, Geochim. et
    Cosmochim. Acta., <u>48</u>, 2219-2229.

58   Tardy, Y. and B. Fritz (1981). An ideal solid solution model for calculating solubility of clay minerals, Clay Minerals, 16, 361-373.

8   Tardy, Y. and R. M. Garrels (1974). A method of estimating the Gibbs energies of formation of layer silicates, Geochim. et Cosmochim. Acta, 38, 1101-1116.

10   Tardy, Y. and R. M. Garrels (1976). Prediction of Gibbs energies of formation - I. Relationships among Gibbs energies of formation of hydroxides, oxides and aqueous ions, Geochim. et Cosmochim. Acta, 40, 1051-1056.

99   Thompson, A. B. (1974). Gibbs energy of aluminous minerals, Contr. Miner. Petrol., 48, 123-136.

45   Vieillard, P., Y. Tardy and D. Nahon (1979). Stability fields of clays and aluminum phosphates: parageneses in lateritic weathering of argillaceous phosphatic sediments, Amer. Miner., 64, 626-634.

4   Vieillard, P. and Y. Tardy (1984). Thermochemical properties of phosphates, Chapter 4 In: Nriagu, J. O. and P. B. Moore, eds., Phosphate Minerals, Springer-Verlag, N.Y., 171-198.

5   Wagman, D. D., W. H. Evans, V. B. Parker, I. Halow, S. M. Bailey and R. H. Schumm (1968-1971). Selected Values of Chemical Thermodynamic Properties, Nat'l. Bureau Standards Tech. Notes 270-4 through 270-7.

31   Wagman, D. D., W. H. Evans, V. B. Parker, R. H. Schumm, I. Halow, S. M. Bailey, K. L. Churney and R. L. Nuttall (1982). The NBS tables of chemical thermodynamic properties: Selected values for inorganic and C_1 and C_2 organic substances in SI units, Journal of Physical and Chemical Reference Data, 11, Supplement No. 2., 392 pp.

116   Wallace, R. M. (1967). Determination of stability constants by donnan membrane equilibrium: the uranyl sulfate complexes, Jour. Phys. Chem., 71, 1271-1276.

41   Weaver, R. M., M. L. Jackson and J. K. Syers (1971). Magnesium and silicon activities in matrix solutions of montmorillonite-containing soils in relation to clay mineral stability, Soil Sci. Soc. Amer. Proc., 35, 823-830.

109   Woods, T. L. and R. M. Garrels (in press). Phase relations of some cupric hydroxy minerals, Econ. Geol.

54   Yang, W. and A. Pourbaix (1981). Effect of chromium and molybdenum on the propagation of localized corrosion of steels, CEBELCOR Rapt. Tech. 262, 141, 42 pp.

16   Zen, E. (1972). Gibbs free energy, enthalpy and entropy of ten rock-forming minerals: Calculations, discrepancies, implications, Amer. Miner., 57, 524-553.

18   Zen, E. (1973). Thermochemical parameters of minerals from oxygen-buffered hydrothermal equilibrium data: Method, applications to annite and almandine, Contr. Miner. Petrol., 39, 65-80.

36   Zverev, V. P. (1982). The Role of Subterranean Water in the Migration of the Chemical Elements (In Russian), Nedra, Moscow, 184 pp.

REFERENCES
Numerical by source number

REFERENCES (numerical by source number)

1   Latimer, W. M. (1952). Oxidation Potentials, Second edition, Prentice-Hall, N.Y., 392 pp.

2   Garrels, R. M. and C. L. Christ (1965). Solutions, Minerals and Equilibria, Freeman, Cooper and Co., San Francisco, Calif., 450 pp.

3   Rossini, F. D., D. D. Wagman, W. H. Evans, S. Levine, and I. Jaffe (1952). Selected Values of Chemical Thermodynamic Properties, Nat'l. Bureau Standards Circ. 500, U. S. Dept. Commerce., Washington, D.C.

4   Vieillard, P. and Y. Tardy (1984). Thermochemical properties of phosphates, Chapter 4 In: Nriagu, J. O. and P. B. Moore, eds., Phosphate Minerals, Springer-Verlag, N.Y., 171-198.

5   Wagman, D. D., W. H. Evans, V. B. Parker, I. Halow, S. M. Bailey and R. H. Schumm (1968-1971). Selected Values of Chemical Thermodynamic Properties, Nat'l. Bureau Standards Tech. Notes 270-4 through 270-7.

6   Naumov, G. B., B. N. Ryzhenko and I. L. Khodakovsky (1974). Handbook of Thermodynamic Data, Nat'l. Tech. Inf. Service, Pb-226, 722/7GA, U. S. Dept. Commerce, 328 pp.

7   Mel'nik, Y. P. (1972). Thermodynamic Constants for the Analysis of Conditions of Formation of Iron Ores (In Russian), Naukova Dumka, Kiev, 196 pp.

8   Tardy, Y. and R. M. Garrels (1974). A method of estimating the Gibbs energies of formation of layer silicates, Geochim. et Cosmochim. Acta, 38, 1101-1116.

9   Stull, D. R. and H. Prophet (1971). JANAF Thermochemical Tables, Second edition, Nat'l. Standard Reference Data Series, Nat'l. Bureau of Standards, 37.

10   Tardy, Y. and R. M. Garrels (1976). Prediction of Gibbs energies of formation. I. Relationships among Gibbs energies of formation of hydroxides, oxides and aqueous ions, Geochim. et Cosmochim. Acta, 40, 1051-1056.

11   Karpov, I. K., A. I. Kiselev and F. A. Letnikov (1971). _Chemical Thermodynamics in Petrology and Geochemistry_, Akademia Nauka, Irkutsk, 385 pp.

12   Pourbaix, M. (1960). Standard free energies of formation at 25°C, CEBELCOR Rapt. 684, Belgian Center for the Study of Corrosion, Brussels, 57 pp.

13   Bricker, O. P., H. W. Nesbitt and W. D. Gunter (1973). The stability of talc, Amer. Miner., _58_, 64-72.

14   Hostetler, P. B. and C. L. Christ (1968). Studies in the system $MgO-SiO_2-CO_2-H_2O$ (I): The activity-product constant of chrysotile, Geochim. et Cosmochim. Acta, _32_, 485-497.

15   Christ, C. L., P. B. Hostetler and R. M. Siebert (1973). Studies in the system $MgO-SiO_2-CO_2-H_2O$ (III): The activity-product constant of sepiolite, Amer. J. Sci., _273_, 65-83.

16   Zen, E. (1972). Gibbs free energy, enthalpy and entropy of ten rock-forming minerals: Calculations, discrepancies, implications, Amer. Miner., _57_, 524-553.

17   Beane, R. E. (1974). Biotite stability in the porphyry copper environment, Econ. Geol., _69_, 241-256.

18   Zen, E. (1973). Thermochemical parameters of minerals from oxygen-buffered hydrothermal equilibrium data: Method, applications to annite and almandine, Contr. Miner. Petrol., _39_, 65-80.

19   Feitknecht, W. and P. W. Schindler (1963). Solubility constants of metal oxides, metal hydroxides and metal hydroxide salts in aqueous solutions, Pure Appl. Chem., _6_, 130-199.

20   Helgeson, H. C. (1969). Thermodynamics of hydrothermal systems at elevated temperatures and pressures, Amer. J. Sci., _267_, 729-804.

21   Robie, R. A., B. S. Hemingway and J. R. Fisher (1978). Thermodynamic properties of minerals and related substances at 298.15 K and 1 bar (10^5 Pascals) pressure and at higher temperatures, U. S. Geol. Survey Bull. 1452, 456 pp.

22   Helgeson, H. C., J. M. Delany, W. H. Nesbitt and D. K. Bird (1978). Summary and critique of the thermodynamic properties of rock-forming minerals, Amer. J. Sci., 278-A, 229 pp.

23   May, H. M., P. A. Helmke and M. L. Jackson (1979). Gibbsite solubility and thermodynamic properties of hydroxy-aluminum ions in aqueous solutions at 25°C, Geochim. et Cosmochim. Acta, 43, 861-868.

24   Karpov, I. K., S. A. Kashik and V. D. Pampura (1968). Thermodynamic Constants for Calculations in Geochemistry and Petrology (In Russian), Izdva Nauka, 143 pp.

25   Lippmann, F. (1979). Stabilitätsbeziehungen der Tonminerale, Neues Jharbuch. Miner., 136, 287-309.

26   Garrels, R. M. (1984). Montmorillonite/illite stability diagrams, Clays and Clay Miner., 32, 161-166.

27   Apps, J. A. (1970). The Stability Field of Analcime, unpub. Ph.D. Thesis, Harvard University, Cambridge, MA.

28   Hemingway, B. S., R. A. Robie and J. A. Kittrick (1978). Revised values for the Gibbs free energy of formation of [$Al(OH)_4^-$ aq], diaspore, boehmite and bayerite at 298.15K and 1 bar, the thermodynamic properties of kaolinite to 800K and 1 bar, and the heats of solution of several gibbsite samples, Geochim. et Cosmochim. Acta, 42, 1533-1543.

29   Parks, G. A. (1972). Free energies of formation and aqueous solubilities of aluminum hydroxides and oxide hydroxides at 25°C, Amer. Miner., 57, 1163-1189.

30   Robie, R. A. and D. R. Waldbaum (1968). Thermodynamic properties of minerals and related substances at 298.15°K (25°C) and one atmosphere (1.013 bars) pressure and at higher temperatures, U. S. Geol. Survey Bull. 1259, 256 pp.

31   Wagman, D. D., W. H. Evans, V. B. Parker, R. H. Schumm, I. Halow, S. M. Bailey, K. L. Churney and R. L. Nuttall (1982). The NBS tables of chemical thermodynamic properties: Selected values for inorganic and C_1 and C_2 organic substances in SI units, Journal of Physical and Chemical Reference Data, 11, Supplement No. 2., 392 pp.

32   Hem, J. D. (1978a). Redox processes at surfaces of manganese oxide and their effects on aqueous metal ions, Chem. Geol., 21, 199–218.

33   Hem, J. D. (1978b). Redox co-precipitation mechanisms of manganese oxides, In: Kavanaugh, M. C. and J. O. Leckie, eds., Particulates in Water, Advances in Chemistry Series, No. 189, 45–72.

34   Lind, C. J. (1978). Polarographic determination of lead hydroxide formation constants at low ionic strength, Envir. Sci. Tech., 12, 1406–1410.

35   Evans, H. T., Jr., and R. M. Garrels (1958). Thermodynamic equilibria of vanadium in aqueous systems as applied to the interpretation of the Colorado Plateau ore deposits, Geochim. et Cosmochim. Acta, 15, 131–149.

36   Zverev, V. P. (1982). The Role of Subterranean Waters in the Migration of the Chemical Elements (In Russian), Nedra, Moscow, 184 pp.

37   Barton, P. B., Jr. and P. M. Bethke (1960). Thermodynamic properties of some synthetic zinc and copper minerals, Amer. J. Sci., 258-A, 21–34.

38   Kittrick, J. A. (1971a). Stability of montmorillonites: I. Belle Fourche and Clay Spur montmorillonites, Soil Sci. Soc. Amer. Proc., 35, 140–145.

39   Kittrick, J. A. (1971b). Stability of montmorillonites: II. Aberdeen montmorillonite, Soil Sci. Soc. Amer. Proc., 35, 820–823.

40   Routson, R. C. and J. A. Kittrick (1971). Illite solubility, Soil Sci. Soc. Amer. Proc., 35, 714–718.

41   Weaver, R. M., M. L. Jackson and J. K. Syers (1971). Magnesium and silicon activities in matrix solutions of montmorillonite-containing soils in relation to clay mineral stability, Soil Sci. Soc. Amer. Proc., 35, 823–830.

42   Mattigod, S. V. and G. Sposito (1978). Improved method for estimating the standard free energies of formation ($\Delta G^{\circ}_{f\ 298.15}$) of smectites, Geochim. et Cosmochim. Acta, 42, 1753–1762.

43   Hemley, J. J., J. W. Montoya, C. L. Christ and P. B. Hostetler (1977a).
      Mineral equilibria in the $MgO-SiO_2-H_2O$ system: I. Talc-chrysotile-
      forsterite-brucite stability relations, Amer. J. Sci., 277, 322-356.

44   Hemley, J. J., J. W. Montoya, D. R. Shaw and R. W. Luce (1977b). Mineral
      equilibria in the $MgO-SiO_2-H_2O$ system: II. Talc-antigorite-forsterite-
      anthophyllite-enstatite stability relations and some geologic implica-
      tions in the system, Amer. J. Sci., 277, 353-383.

45   Vieillard, P., Y. Tardy and D. Nahon (1979). Stability fields of clays
      and aluminum phosphates: parageneses in lateritic weathering of
      argillaceous phosphatic sediments, Amer. Miner., 64, 626-634.

46   Huang, W. H. and W. D. Keller (1973). Gibbs free energies of formation
      calculated from dissolution data using specific mineral analyses. III.
      Clay minerals, Amer. Miner., 58, 1023-1028.

47   Hemingway, B. S. and R. A. Robie (1977a). Enthalpies of formation
      of low albite ($NaAlSi_3O_8$), gibbsite ($Al(OH)_3$) and $NaAlO_2$; revised
      values for $\Delta H^\circ_{f, 298}$ and $\Delta G^\circ_{f, 298}$ of some aluminosilicate minerals,
      J. Research U.'S. Geol. Survey, 5, 413-429.

48   Hemley, J. J., P. B. Hostetler, A. J. Gude and W. T. Mountjoy (1969).
      Some stability relations of alunite, Econ. Geol., 64, 599-612.

49   Reeseman, A. L. (1974). Aqueous dissolution studies of illite under
      ambient conditions, Clays and Clay Miner., 22, 443-454.

50   Sangameshwar, S. R. and H. L. Barnes (1983). Supergene processes in zinc-
      lead-silver sulfide ores in carbonates, Econ. Geol., 78, 1379-1397.

51   Charlu, T. V., R. C. Newton and O. J. Kleppa (1975). Enthalpies of
      formation at 970K of compounds in the system $MgO-Al_2O_3-SiO_2$ from high
      temperature solution calorimetry, Geochim. et Cosmochim. Acta, 39,
      1487-1497.

52   Hemingway, B. S. and R. A. Robie (1977b). The entropy and Gibbs free
      energy of formation of the aluminum ion, Geochim. et Cosmochim. Acta,
      41, 1402-1404.

53 Saxena, S. K. (1976). Entropy estimates for some silicates at 298°K from molar volumes, Science, $\underline{193}$, 1241-1242.

54 Yang, W. and A. Pourbaix (1981). Effect of chromium and molybdenum on the propagation of localized corrosion of steels, CEBELCOR Rapt. Tech. 262, $\underline{141}$, 42 pp.

55 Pourbaix, M. and X. Yang (1981). Chemical and electrochemical equilibria in the presence of a gaseous phase. 5. -oxygen-hydrogen-iron, CEBELCOR Rapt. Tech. 1401, $\underline{260}$, 116 pp.

56 Bricker, O. P. (1965). Some stability relationships in the system $Mn-O_2-H_2O$ at 25°C and 1 atm. total pressure, Amer. Miner., $\underline{50}$, 1296-1354.

57 Hem, J. D. and C. E. Roberson (1967). Form and stability of aluminum hydroxide complexes in dilute solution, U. S. Geol. Survey Water-Supply Paper 1827-A, A1-A55.

58 Tardy, Y. and B. Fritz (1981). An ideal solid solution model for calculating solubility of clay minerals, Clay Minerals, $\underline{16}$, 361-373.

59 Johnson, G. K., H. E. Flotow and P. A. G. O'Hare (1983). Thermodynamic studies of zeolites: natrolite, mesolite and scolecite, Amer. Miner., $\underline{68}$, 1134-1145.

60 Leussing, D. L. and I. M. Kolthoff (1953). The solubility product of ferrous hydroxide and the ionization of the aquo-ferrous ion, J. Amer. Chem. Soc., $\underline{75}$, 2746.

61 Morgan, J. J. (1967). Chemical equilibria and kinetic properties of manganese in natural waters, In: Faust, S. D. and J. V. Hunter, eds., Principles and Applications of Water Chemistry, John Wiley and Sons, New York, 561-624.

62 Pourbaix, M. (1963). Atlas d'equilibres electrochimiques a 25°C, Gauthier-Villars, Paris, 644 pp.

63 Hemley, J. J., J. W. Montoya, J. W. Marinenko and R. W. Luce (1980). Equilibria in the system $Al_2O_3-SiO_2-H_2O$ and some general implications for alteration/mineralization processes, Econ. Geol., $\underline{75}$, 210-228.

64   Hem, J. D., C. E. Roberson and C. J. Lind (1985). Thermodynamic stability
      stability of CoOOH and its coprecipitation with manganese, Geochim. et
      Cosmochim. Acta, 49, 801-810.

65   Garrels, R. M., M. E. Thompson and R. Siever (1960). Stability of some
      carbonates at 25°C and 1 atmosphere total pressure, Amer. J. Sci., 258,
      402-418.

66   Robie, R. A. and B. S. Hemingway (1984). Entropies of kyanite, andalus-
      ite and sillimanite: additional constraints on the pressure and tem-
      perature of the Al_2SiO_5 triple point, Amer. Miner., 69, 298-306.

67   Hemingway, B. S. and R. A. Robie (1984). Heat capacity and thermodynamic
      functions for gehlenite and staurolite: with comments on the Schottky
      anomaly in the heat capacity of staurolite, Amer. Miner., 69, 307-318.

68   Raymahashay, B. C. (1968). A geochemical study of rock alteration by hot
      springs in the Paint Pot Hill area, Yellowstone Park, Geochim. et
      Cosmochim. Acta, 32, 499-522.

69   Bricker, O. P. (1969). Stability constants and Gibbs free energies of
      formation of magadiite and kenyaite, Amer. Miner., 54, 1026-1033.

70   Bullwinkel, E. P. (1954). The chemistry of uranium in carbonate solu-
      tions, U. S. Atomic Energy Commission, Raw Materials Division RMO,
      2614, 59 pp.

71   Couturier, Y., G. Michard and G. Sarazin (1984). Constantes de formation
      des complexes hydroxydes de l'aluminium en solution aqueuse de 20 a
      70°C, Geochim. et Cosmochim. Acta, 48, 649-660.

72   Harvie, C. E., N. Moller and J. H. Weare (1984). The prediction of
      mineral solubilities in natural waters: The $Na-K-Mg-Ca-H-Cl-SO_4-OH-$
      $HCO_3-CO_3-CO_2-H_2O$ system to high ionic strengths at 25°C, Geochim. et
      Cosmochim. Acta, 48, 723-751.

73   Haas, H. and M. J. Holdaway (1973). Equilibria in the system $Al_2O_3-SiO_2-$
      H_2O involving the stability limits of pyrophyllite, and thermodynamic^2
      data of pyrophyllite, Amer. J. Sci., 273, 449-464.

74 Haas, H. and M. J. Holdaway (1974). Equilibria in the system Al_2O_3-SiO_2-H_2O involving the stability limits of pyrophyllite, and thermodynamic data of pyrophyllite: Additional data, Amer. J. Sci., 274, 825-828.

75 Bird, G. W. and G. M. Anderson (1973). The free energy of formation of magnesian cordierite and phlogopite, Amer. J. Sci., 273, 84-91.

76 Christ, C. L., P. B. Hostetler and R. M. Siebert (1974). Stabilities of calcite and aragonite, J. Research U. S. Geol. Survey, 2, 175-184.

77 Karpov, I. K., A. I. Kiselev and F. A. Letnikov (1976). Modeling of natural mineral assemblages with an IBM, Nedra, Moscow, 256 pp.

78 Reeseman, A. L. and W. D. Keller (1968). Aqueous solubility studies of high-alumina and clay minerals, Amer. Miner., 53, 929-942.

79 Silman, J. F. B. (1958). The stabilities of some oxidized copper minerals in aqueous solutions at 25°C and 1 atmosphere total pressure, unpub. Ph.D. thesis, Harvard University, Cambridge, MA, 98 pp.

80 Helgeson, H. C. (1983 and 1984). Supcrt Update Notices.

81 Kelley, K. K. (1937). Contributions to the data on theoretical metallurgy. VII. The thermodynamic properties of sulfur and its inorganic compounds, U. S. Bureau Mines Bull. 406, 154 pp.

82 Garrels, R. M. (1957). Some free energy values from geologic relations, Amer. Miner., 42, 780-792.

83 King, E. G., R. Barany, W. W. Weller and L. B. Pankratz (1967). Thermodynamic properties of forsterite and serpentine, U. S. Bureau Mines Rept. Inv. 6962, 19 pp.

84 Chernosky, J. V., Jr. (1974). The upper stability of clinochlore at low pressure and the free energy of formation of Mg-cordierite, Amer. Miner., 59, 496-507.

85 Hepler, L. G. and G. Olofsson (1975). Mercury: Thermodynamic Properties, Chemical Equilibria, and Standard Potentials, Chem. Rev., 75, 585-602.

86 Kittrick, J. A. (1970). Precipitation of kaolinite at 25°C and 1 atm.,
    Clays and Clay Miner., 18, 261-267.

87 Polzer, W. L. and J. D. Hem (1965). The dissolution of kaolinite,
    J. Geophys. Research, 70, 6233-6240.

88 Saegusa, F. (1950). Science reports, Tokoku Univ., 1st Series, 34, 104.

89 Raupach, M. (1963a). Solubility of simple aluminum compounds expected in
    soils. I. Hydroxides and oxyhydroxides, Australian J. Soil Research, 1,
    28-35.

90 Kittrick, J. A. (1966). The free energy of formation of gibbsite and
    $Al(OH)_4^-$ from solubility measurements, Soil Sci. Soc. Amer. Proc., 30,
    595-598.

91 Barton, M. D., H. T. Haselton, Jr., B. S. Hemingway, O. J. Kleppa
    and R. A. Robie (1982). The thermodynamic properties of fluor-
    topaz, Amer. Miner., 67, 350-355.

92 Barton, M. D. (1982). The thermodynamic properties of topaz solid solu-
    tions and some petrologic applications, Amer. Miner., 67, 956-974.

93 Pourbaix, M. (1973). Lectures on Electrochemical Corrosion, Plenum
    Press, New York and London, 336 pp.

94 Langmuir, D. and A. C. Riese (1985). The thermodynamic properties of
    radium, Geochim. et Cosmochim. Acta, 49, 1593-1601.

95 Reinert, M. (1965). Über die Bestimmung der Löslichkeit schwerlöslicher
    Salze mit basischen Anionen, unpub. Ph. D. Thesis, University of Bern.

96 Schindler, P., H. Althaus, F. Hofer and W. Minder (1965). Löslichkeits-
    produkte von Zinkoxid, kupferhydroxid und kupferoxid in Abhängigkeit
    von Teilehengrösse und molarer Oberfläche. Ein Beitrag zur Thermody-
    namik von Grenzflächen fest-flüssig. Helv. Chim. Acta., 48, 1204-1215.

97 Sillen, L. G. and A. E. Martell (1964). Stability Constants of Metal-
    Ion Complexes, Spec. Publ. 17, The Chem. Soc., Burlington House,
    London, 754 pp.

98   Richet, P., Y. Bottinga, L. Denielou, J. P. Petitet and C. Tequi (1982). Thermodynamic properties of quartz, cristobalite and amorphous SiO_2: drop calorimetry measurements between 1000 and 1800K and a review from 0 to 2000K, Geochim. et Cosmochim. Acta, 46, 2639-2658.

99   Thompson, A. B. (1974). Gibbs energy of aluminous minerals, Contr. Miner. Petrol., 48, 123-136.

100   Crerar, D. A. and H. L. Barnes (1974). Deposition of deep-sea manganese nodules, Geochim. et Cosmochim. Acta, 38, 279-300.

101   Rickard, D. T. and J. O. Nriagu (1978). Aqueous environmental chemistry of lead, In: Nriagu, J. O., ed., The Biogeochemistry of Lead in the Environment, Elsevier - North Holland Biomedical Press, New York, 219-284.

102   Mann, A. W. and R. L. Deutscher (1980). Solution geochemistry of lead and zinc in water containing carbonate, sulfate, and chloride ions, Chem. Geol., 29, 293-311.

103   Symes, J. L. and D. R. Kester (1984). Thermodynamic stability studies of the basic copper carbonate mineral, malachite, Geochim. et Cosmochim. Acta., 48, 2219-2229.

104   Rickard, D. T. (1970). The chemistry of copper in natural aqueous solutions, Acta Universitatis Stockholmiensis, Stockholm Contributions in Geology, 23:1, 64pp.

105   Perkins, D., III, E. J. Essene, E. F. Westrum, Jr. and V. J. Wall (1979). New thermodynamic data for diaspore and their application to the system Al_2O_3-SiO_2-H_2O, Amer. Miner., 64, 1080-1090.

106   Robie, R. A., H. T. Haselton, Jr. and B. S. Hemingway (1984). Heat capacities and entropies of rhodochrosite ($MnCO_3$) and siderite ($FeCO_3$) between 5 and 600 K, Amer. Miner., 69, 349-357.

107   Newberg, Donald W. (1967). Geochemical implications of chrysocolla-bearing alluvial gravels, Econ. Geol., 62, 932-956.

108 Robinson, G. R., Jr., J. L. Haas, Jr., C. M. Schafer and F. T.
    Haselton, Jr. (1982). Thermodynamic and thermophysical properties
    of selected phases in the $MgO-SiO_2-H_2O-CO_2$, $CaO-Al_2O_3-SiO_2-H_2O-$
    CO_2, and $Fe-FeO-Fe_2O_3-SiO_2$ chemical systems, with special empha-
    sis on the properties of basalts and their mineral components,
    U. S. Geol. Survey Open-file Report 83-79, 429 pp.

109 Woods, T. L. and R. M. Garrels (in press). Phase relations of some
    cupric hydroxy minerals, Econ. Geol.

110 Robie, R. A., B. S. Hemingway, J. Ito and K. M. Krupka (1984). Heat
    capacity and entropy of Ni_2SiO_4-olivine from 5 to 1000K and heat
    capacity of Co_2SiO_4 from 360 to 1000K, Amer. Miner., 69, 1096-1101.

111 Fuger, J. and F. L. Oetting (1976). The chemical thermodynamics
    of actinide elements and compounds. Pt. 2, The actinide aqueous
    ions, Internat. Atomic Energy Agency, Vienna, 65 pp.

112 Baes, C. F., Jr. and R. E. Mesmer (1976). The Hydrolysis of Cations,
    John Wiley and Sons, New York, 489 pp.

113 Nikitin, A. A., Z. I. Sergeyeva, I. L. Khodakovsky and G. B. Naumov
    (1972). Uranyl hydrolysis in a hydrothermal region, Geokhimiya, 3,
    297-307.

114 O'Cinneide, S., J. P. Scanlan and M. J. Hynes (1972). Overall stability
    constant of the complex $UO_2(CO_3)_3^{-4}$, Chem. Indus. (London), 50, 340.

115 Sergeyeva, E. I., A. A. Nikitin, I. L. Khodakovsky and G. B. Naumov
    (1972). Experimental investigation in the system $UO_3-CO_2-H_2O$ in
    25-200°C temperature interval, Geochemistry Internat., 9, 900-910.

116 Wallace, R. M. (1967). Determination of stability constants by donnan
    membrane equilibrium: the uranyl sulfate complexes, Jour. Phys. Chem.,
    71, 1271-1276.

117 Langmuir, D. and K. Applin (1977). Refinement of the thermodynamic
    properties of uranium minerals and dissolved species with applications
    to the chemistry of ground waters in sandstone-type uranium deposits,
    In: Campbell, J. A., ed., Short Papers of the U. S. Geol. Survey
    Uranium-Thorium Symp., U. S. Geol. Survey Circ. 753, 57-66.

118 Johnson, G. K., H. E. Flotow, P. A. G. O'Hare and W. S. Wise (1985).
     Thermodynamic studies of zeolites: heulandite, Amer. Miner., $\underline{70}$,
     1065-1071.

119 Babushkin, V. I., G. M. Matveyev and O. P. Mchedlov-Petrossyan (1985).
     Thermodynamics of Silicates, Springer-Verlag, Berlin, 459 pp.

120 Nesbitt, H. W. (1977). Estimation of the thermodynamic properties of Na-
     Ca- and Mg- beidellites, Canadian Miner., $\underline{15}$, 22-30.

121 Hemingway, B. S., J. L. Haas, Jr. and G. R. Robinson, Jr. (1982).
     Thermodynamic properties of selected minerals in the system Al_2O_3-
     CaO-SiO_2-H_2O at 298.15 K and 1 bar (10^5 pascals) pressure and at
     higher temperatures, U. S. Geol. Survey Bull. 1544, 70 pp.